Algorithm Engineering for Integral and Dynamic Problems

PARALLEL PROCESSING

A series edited by G.M. Megson, University of Reading, UK

Volume 1
Piecewise Regular Arrays: Application-specific computations
Toomas P. Plaks

Volume 2
Algorithm Engineering for Integral and Dynamic Problems
Lucia Rapanotti

This book is part of a series. The publisher will accept continuation orders which may be cancelled at any time and which provide for automatic billing and shipping of each title in the series upon publication. Please write for details.

Algorithm Engineering for Integral and Dynamic Problems

Lucia Rapanotti

The Open University, Milton Keynes, UK

GORDON AND BREACH SCIENCE PUBLISHERS

Australia • Canada • France • Germany • India • Japan
Luxembourg • Malaysia • The Netherlands • Russia
Singapore • Switzerland

Amsteldijk 166
1st Floor
1079 LH Amsterdam
The Netherlands

British Library Cataloguing in Publication Data

ISBN 90-5699-328-3
ISSN 1028-8937

To Jon and Gabriele,
my family, my life.

Contents

List of Figures

List of Tables

Preface

Algorithm specific computing provides architectural solutions for the efficient execution of classes of computation-intensive algorithms. This is achieved by exposing the algorithm's inherent parallelism and defining a parallel machine that can execute it. As the number of algorithms which require efficient execution grows, algorithm engineers strive to provide effective architectural solutions to wider and wider classes of problems.

Over the past fifteen years, systolic and regular arrays have proven to be effective architectures for the efficient execution of many computation-intensive algorithms. Synthesis techniques for regular arrays assist algorithm engineers by providing a *disciplined* and *well-founded* approach to the engineering of classes of parallel algorithms: disciplined, as a design process is defined, which consists of the application of a number of well-defined transformation; well-founded, as both algorithm specifications and their transformations are expressed in a mathematical notation and equipped with a mathematical semantics.

The mathematical theory underlying these synthesis techniques is that of affine Euclidean geometry with embedded lattice spaces. This is a powerful theory not just for the denotation of parallel algorithms, but also because it comes with computational solutions which can be exploited in the engineering of regular processor arrays. In particular, parallel algorithms are treated within this theory by representing their data dependencies as vectors in Euclidean lattice spaces. Based on this mathematical embedding, a number of properties can be investigated, and transformations applied.

In general, current synthesis techniques are limited in that they have been successfully applied only to so-called *affine* algorithms. These are algorithms whose defining data dependencies can be expressed as affine transformations in the underlying mathematical theory. The work presented in this book aims at widening the applicability of standard synthesis techniques to more general classes of algorithms.

The major contributions reported in this book are the characterisation of classes of integral and dynamic algorithms, and the provision of techniques for their systematic treatment within the framework of established synthesis methods. The basic idea is to transform the initial algorithm specification into a specification with data dependencies of increased regularity so that corresponding regular arrays can be obtained by a direct application of the standard mapping techniques. We have applied our techniques to well-known algorithms in the literature, and many case studies are presented in the book.

Audience

This book is aimed at researchers and practitioners in the field of algorithm engineering. In this book they can find a comprehensive treatment of algorithm transformations for the derivation of regular processor arrays.

We assume that the reader is familiar with the basic concepts underpinning parallel algorithms and architectures, as well as with the mathematical theories of linear and affine algebra as presented in undergraduate Computing degrees. The reader will also benefit from some familiarity with the systolic and regular processor arrays literature.

This book will be of value to doctoral students who are interested in the synthesis of regular algorithms. The introductory Chapters 1 and 2 can also be used as reference material by teachers on advanced courses on parallel processing.

Acknowledgments

I would like to thank Professor Graham M. Megson, as editor of this series and my PhD supervisor, for suggesting this subject of research and for his invaluable guidance and advice throughout my doctoral work.

I would also like to thank Professor Fiorella De Cindio of *Università degli Studi* of Milan, who first gave me the opportunity to pursue a career in computing research and education.

I am grateful to colleagues, past and present, in the Computing Departments of the University of Newcastle upon Tyne and the Open University for providing a friendly and stimulating working environment. In particular, I would like to thank Shirley Craig and Trevor Kirby, who have been most helpful on so many occasions.

Last, but not least, a special thank you goes to my husband, Jon Hall, for the patience, loving support and encouragement over the years. Our lively discussions over the content of this book have helped me to greatly improve the quality of my work.

Chapter 1

Introduction

Parallel and distributed computing represent an important branch of Computing Science. Many of today's computer applications require a great computing power at very high speed. Higher performance for these applications can be achieved through parallel processing, that is by allowing the concurrent and cooperative execution of their computations. Parallel and distributed computing is a necessary and effective alternative to building faster sequential computers: necessary as hardware components are approaching their technological limits as for the level of circuit integration and the speed of signal propagation; effective as significant "speed-ups" can be obtained for many applications of interest. Also, technological advances have made it possible for both commercial and research parallel computers and distributed systems to be widely available at a relatively low cost. Today's research in parallel and distributed computing spans from architectural and algorithmic issues, to programming languages and compiler technology, to theoretical models and complexity theory. A comprehensive overview of these issues is outside the scope of this book, but good sources of information can be found, for instance, in [Cri97, Zom95, Kri89, HwBr85].

A substantial part of the research in parallel computing has been devoted to the study of parallel machines which are *algorithm specific* [Sn-et-al85]: algorithm specific machines are machines whose architectures are specialised to provide efficient solutions for classes of problems which share a common solution method. Hence algorithm specific machines provide a trade-off in which improved performance is gained at the expense of generality. This book deals with a particular aspect of algorithm specific computing, that of the synthesis of *regular processor arrays*, or *regular arrays*.

1.1   Algorithm Specialisation

The basic objective of algorithm specialisation is the definition of a parallel machine for the efficient execution of an algorithm. Efficiency is obtained by an optimal exploitation of the *structural properties* of the algorithm within a number of *design constraints*. The structural properties of an algorithm are those related, for instance, to its sub-tasks, their generation and data communications. Architectural specialisation is based on such properties. Design constraints are those related, for instance, to processing speed, physical size, or accuracy of the results. Design constraints are typical of the application area of an algorithm, or imposed by the available physical resources.

A methodological approach to algorithm specialisation considers the definition of an algorithm specific machine as a design process including methods for the formal description and transformation of an algorithm, and optimisation strategies for the optimal exploitation of its structural properties within a given set of constraints.

We may classify methods for algorithm specialisation into two broad categories. On the one hand, we find general design strategies[1], including *divide and conquer, branch and bound, dynamic programming, search and traversal methods*, or *backtracking* [Kri89]. Given a problem, the designer adopts a number of these strategies until a satisfactory algorithm specific machine is defined. In general, in this approach to design, the designer has to be an expert in order to apply the various design strategies optimally for the particular problem. Usually, little automatic support can be provided because of the *ad hoc* transformations which are involved. Design strategies, however, have the benefit of being generally applicable and likely to produce optimal solutions given a problem's requirements.

On the other hand, there are methods that focus on a particular model of parallel computation and try and identify those algorithms which can be efficiently executed within such a model. This viewpoint has led to the development of parallel compilation techniques and systematic synthesis of regular arrays. The emphasis, in this case, is on the syntactic characterisation of algorithms and their systematic manipulation by formal transformations. In this approach a prominent rôle is given to the development of automatic support, so that the transformations are not only systematic, but also largely mechanised. Because of the relevance given to tool support, in

---

[1]Most of these strategies were actually developed independently and prior to parallel computing, motivated by the desire of devising optimal algorithms for applications of interest.

the design process a lower level of expertise is expected from the designer. However, there is a loss of generality, as only algorithms which conform to a particular syntax can be treated. Also, sub-optimal solutions are likely to be derived, as optimisation strategies are formulated in general terms for classes of problems instead of being targeted to particular applications.

In this work we take this second approach and aim at developing design methods for the systematic synthesis of regular arrays.

1.2 Regular Array Synthesis

The type of parallelism we address in this work is that characteristic of *regular arrays*, i.e., synchronous regularly connected networks of processors. We will refer to this form of parallelism as *regular parallelism*. Regular parallelism is *synchronous* and *deterministic* in that data are supposed to be transferred through the processor network at regular, consecutive and specified instants of time. Regular parallelism is *massive* in that the number of processors in the network is assumed to be of the same order of the size parameters of the problem, with each processor performing simple operations, corresponding to the basic computations of the algorithm.

Historically, the development of regular arrays and their synthesis techniques can be related to the advent of VLSI design and fabrication techniques, and was initiated, in the late 70s, by Kung and Leiserson [KuLe80] with the introduction of *systolic arrays*. Regular arrays have developed from systolic arrays by relaxing some of the initial constraints on the topology of the network and the complexity of the processing elements (such as strictly neighbour connections, or bit and word level operations). Also, in time regular arrays have evolved from being seen as particular types of hardware components to being considered as special types of synchronous parallel programs, hence they have been adopted as a model of parallel computation. A brief survey of the development of regular array synthesis with the relevant references is given in Chapter 2.

This development of the subject has brought regular array synthesis closer to parallel compilation techniques for Fortran-like programs, and, in particular, the automatic parallelisation of nested for-loops [Wol89]. Besides, together with the traditional exploitation of parallelism for vector processors and shared memory machines, parallel compilation techniques have also evolved to include techniques for distributed memory architectures [OBo93]. The combined effect of these developments have resulted in an even stronger bond between parallel compilation and regular array synthesis [Me-et-al95b].

1.2.1   Algorithm Specification

Synthesis methods for regular arrays are based on the systematic manipulation of *algorithm descriptions* or *specifications*[2]. A specification conforms to
a formal syntax and corresponds to a functional description of the algorithm.
Restricted forms of imperative nested for-loops or recurrence equations are
usually admitted as a specification (no universally adopted syntax exists in
the literature). For example, an algorithm which computes the first $n + 1$
entries of the Fibonacci sequence could be specified as the code segment:

$F(0) := 1;$
$F(1) := 1;$
for $i := 2$ to n do
$\quad F(i) := F(i-1) + F(i-2);$

or the recurrence equation:

$$F(i) = \begin{cases} 1 & i = 0, 1 \\ F(i-1) + F(i-2) & i = 2, \ldots, n \end{cases}$$

The specification does not contain any explicit directive for the parallel
execution of the computations of the algorithm, because, in principle, the
specifier should not be concerned with the model of computation adopted.

Although, it could be argued that the use of imperative code may imply a
degree of awareness of some model of computation by the algorithm designer,
its traditional use is for syntactic description only. Indeed, the subset of
imperative code used corresponds to so-called *single assignment code*, which
means that assignment statements cannot have a destructive effect on the
value of an already assigned variable; in other words, each variable is assigned
exactly once during the execution of the algorithm [Lis89]. To a certain
extent, the use of single assignment code allows the specifier to adopt a
(perhaps more familiar) imperative programming style, while a functional
interpretation of the specification is assumed by the method. The reader
interested in the debate imperative *vs.* functional is referred to the famous
article by Backus in [Bac78].

---

[2]In the following, we will often use the terms algorithm and specification indifferently
to indicate the formal description of an algorithm.

Our main concern in this work is the development of methods for the systematic synthesis of regular arrays, and a purely functional approach to the specification of algorithms will be adopted. We will, however, use nested for-loops in examples and illustrations, so that the reader less accustomed to mathematical notations may gain an intuition of the effects of the formal transformations. Work on how to convert imperative nested for-loops into functional descriptions exists in the literature. See, for instance, [BuDe88, Lis89].

1.2.2 Parallelism and Data Dependence Relations

While there are no directives for concurrency or data communication, the algorithm specification contains a description of the *data dependence relations* between the computations of the specified algorithm.

A data dependence relation introduces a sequentiality constraint by expressing that the execution of some computations relies on data generated by other computations. For instance, if we consider the specification of the Fibonacci sequence of the previous section, the evaluation of F at each i depends on the evaluations of F both at $i - 1$ and $i - 2$. Hence, the entries of F have to be computed sequentially for i from 2 to n.

Fortunately, strictly sequential computations do not characterise all algorithms. Indeed, the source of the exploitable parallelism of a specification is represented by the sets of its computations which are not related under any data dependence relation. For instance, consider the following segment of code, which computes a vector C whose entries are the sums of two adjacent entries of a given vector A:

```
for i := 2 to n do
    C[i] := A[i] + A[i − 1];
```

The corresponding equation is:

$$C(i) = A(i) + A(i - 1) \qquad i = 2, \ldots, n$$

Data dependence relations can be established between C and A. In particular, for all i, the evaluation of C at i depends on the values of A both at i and $i - 1$. On the other hand, there is no data dependence between the computations of C for different values of the index i. Therefore, the entries of C may be computed in parallel.

That the ordering of the operations of a program is based on the needs of data (instead of being specified by the programmer) is the basic principle of *data flow computing* [Ada68, Ada70, DeWe77, Den80]. The analysis of the data dependencies of a program has a prominent rôle in regular array synthesis, and more generally in the exploitation of algorithm parallelism and parallel compilation techniques, and a number of tools have been developed for the representation of data dependence relations (mainly based on graph theory [Car79]). A basic tool, which we will use extensively in this work, is the so-called *data dependence graph*, which provides a graphical representation of the data dependence relations of an algorithm. Nodes in a data dependence graph represent computations, with arcs representing their data interdependencies.

1.2.3   Algorithm Specialisation and Space-Time Mapping

By decomposing an algorithm into basic computations and their interdependencies, the maximal inherent parallelism of the algorithm is uncovered. In particular, if the corresponding data dependence graph defines a partial order, its sets of incomparable nodes correspond to computations that can be executed in parallel, while its arcs indicate the data communications required between computations. In being a graphical representation, the data dependence graph allows for structural and topological considerations. In particular, a processor network whose topology matches the data dependence graph may provide a specialised parallel machine for the execution of the algorithm. The correspondence couples processors of the network with nodes of the data dependence graph (computations) and communication links with arcs of the data dependence graph (data communications).

More complex correspondences can be established between a data dependence graph and a network of processors, for instance by associating several computations to the same processor. In this case, care should be taken that the partial order induced by the data dependencies is reflected by the ordering of execution of the computations in the processor network.

Therefore, in regular array synthesis we can consider algorithmic specialisation as a mapping of the data dependence graph of the algorithm onto a processor network. This mapping is usually known as a *space-time mapping*, as it can be seen as characterised both by a spatial and a temporal component. The spatial component defines a correspondence between computations and processors. The temporal component indicates the ordering of the computations at each processor.

1.2.4   The Rôle of Regularity

A space-time mapping is straightforward only if data dependence graph and processor network share a similar topology. When the target processor network is a regular array, we are left with the problem of characterising regular data dependence graphs and the type of algorithms to which they correspond. In other words, we need to identify those properties of an algorithm which yield regular data dependence graphs and, hence, regular array designs.

Regular array synthesis has developed from the realisation that particular forms of nested for-loops and recurrence equations correspond to regular data dependence graphs. They are characteristic of so-called *uniform* problems [Ka-et-al67]. The term uniform refers to characteristics of the data dependencies of the problem. Our description of the Fibonacci sequence is an example of a uniform algorithm. In particular, for all i, the computations of F are characterised by the data dependence of $F(i)$ on $F(i-1)$ and $F(i-2)$. In terms of data dependence graph, the same arcs, from $i-1$ to i and from $i-2$ to i, are uniformly replicated at each node of the graph (which is then highly regular). Uniform problems can be easily recognised by considering the form of the index expressions of their variables. If the indices are seen as representing the axes of a Euclidean space, uniform index expressions correspond to translations in that space.

From a designer's point of view, uniform recurrences offer a limited abstraction power for the specification of algorithms. The expression of a generic algorithm as uniform nested for-loops or uniform recurrences requires considerable effort and expertise from the designer, often involving several manipulations of (from a human's point of view) more natural expressions of the algorithm. It also implies a high level of awareness of the underlying model of computation.

In order to increase the abstraction power of the languages for algorithm specification, so-called *affine* problems have been considered [Mol83, RaFu90]. This type of specifications are characterised by index expressions which are affine expressions of the indices, that is the combination of linear expressions and translations [Ner63, Roc70, Sch86]. An example of an affine specification is the following set of nested for-loops for the computation of the product of two $n \times n$ matrices A and B:

```
for i := 1 to n do
  for j := 1 to n do
```

```
begin
    C(i, j, 0) := 0;
    for k := 1 to n do
        C(i, j, k) := C(i, j, k − 1) + A(i, k) * B(k, j);
end;
```

The recurrence formulation of the same algorithm is:

$$C(i, j, k) = \begin{cases} 0 & i, j = 1, \ldots, n, \ k = 0 \\ C(i, j, k-1) + A(i, k) * B(k, j) & i, j, k = 1, \ldots, n \end{cases}$$

Note that the single assignment form of the code requires a third index for the accumulation variable C. The entries of the result matrix correspond to the values $C(i, j, n)$ for all combinations of i and j. Each computation $C(i, j, k)$ depends on the computation of the values $C(i, j, k − 1)$, $A(i, k)$ and $B(k, j)$, and the relation between each pair of index expressions can be described as an affine mapping.

As uniform index expressions define translations and affine index expressions combine translations and linear transformations, then a uniform data dependence relation is a particular case of affine data dependence relation. Hence, the power of abstraction of affine specifications is higher than that of uniform specifications. The gain in abstraction is, however, counterbalanced by a loss of regularity of the data dependencies.

1.2.5 Enforcing Regularity

The increase of abstraction power from uniform to affine can be usefully exploited only if it is accompanied by the provision of systematic transformations of an affine description into a uniform algorithm. These transformations have, indeed, been the subject of study by several authors and the problem of making an affine problem uniform is now well-understood [FoMo84, RaFu87, Raj89, WoDe92, QuVa89]. As the effect of these transformations is that of enforcing regularity, we call them *regularising transformations*. This is not a standard terminology. Other names which have appeared in the literature include *uniformisation* and *localisation*. We feel, however, that regularity carries a more general meaning than the others – in fact, uniform does not necessarily mean local, while locality does not imply uniformity.

With the introduction of regularising transformations, algorithm specialisation through regular array synthesis can be logically outlined as in the

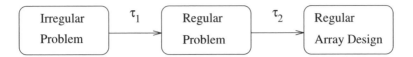

Fig. 1.1. Outline of Regular Array Synthesis.

diagram of Fig. 1.1. Initially, a non-regular specification is provided. This is transformed into a regular algorithm (transformation τ_1) by the application of regularising transformations, and subsequently mapped onto a regular array (transformation τ_2) through a mapping of its regular data dependence graph.

Although in the diagram regularisation has been represented as a mapping from non-regular to regular algorithms, it should be represented more faithfully as a transformational process iterating on non-regular specifications, and which terminates with a regular algorithm. This is because, in general, regularisation techniques apply to single non-uniform data dependencies, rather than to a specification as a whole. For instance, the matrix product example of the previous section would be transformed by classic regularisation techniques into a uniform algorithm by transforming separately the data dependence relations between $C(i,j,k)$ and $A(i,k)$ and between $C(i,j,k)$ and $B(k,j)$ (the relation between $C(i,j,k)$ and $C(i,j,k-1)$ is already uniform).

The fact that regularising transformations apply to single data dependence relations has a significant impact on the simplicity of their definition and software implementation. Indeed, regularising transformations have also to be consistent with optimisation strategies which apply to the problem as a whole.

The balance between scope of application, complexity and optimality of the transformations plays an important rôle in the design process. While fully automated optimisation is not supported in regular array synthesis as yet (mainly because an exhaustive search of the solution space of the problem would not be computationally feasible), a possibility is offered to the designer to choose an optimisation strategy and perform some basic transformations of the specification automatically.

1.3   From Affine to Integral Problems

Although affine data dependencies characterise a large number of problems, they still impose severe restrictions on the specification of algorithms. For instance, in the set of nested loops below, variable P accumulates the values of m entries of variable Q, where we assume that Q had been suitably initialised before these for-loops are executed. The m entries of Q are determined by the index expression $i * j$. Unfortunately this index expression is not an affine expression, hence we cannot apply current synthesis methods.

```
for i := 1 to n do
   begin
      P(i, 0) := 0;
      for j := 1 to m do
         P(i, j) := P(i, j − 1) + Q(i * j);
   end;
```

The limitations of the present techniques are mainly due to the choice of the underlying mathematical model for regular array synthesis, that of linear algebra and affine geometry [Ner63, Roc70, Sch86]. Linear and affine transformations as well as convex polyhedral sets are the key concepts which are exploited by synthesis techniques both from a theoretical and an applicative point of view. Hence data dependencies which are defined through arbitrary index expressions fall outside the scope of the established synthesis techniques.

That synthesis techniques are not applicable does not imply that regular parallel solutions are not possible for non-affine problems. It simply means that the techniques are not powerful enough to provide such solutions systematically. Indeed, *ad hoc* regular arrays for non-affine problems have been proposed in the literature (see, e.g., [Eva91]).

One of the main objectives of the work presented in this book is the characterisation of classes of non-affine problems and the provision of regularising techniques for their systematic transformation into uniform specifications. The characterisation that we will provide yields the definition of *integral problems* as problems whose index expressions are integer mappings of the indices.

1.4 From Static to Dynamic Problems

Both affine and integral problems share the property that their data dependencies are entirely specified by the algorithm description. Hence, a *static* or *compile-time* analysis of those data dependencies is possible, as well as their representation as a data dependence graph. There are problems, however, that do not share this property. For instance, consider the code segment below (obtained through a small modification of the code segment of the previous section):

```
for i := 1 to n do
   read(G(i));
for i := 1 to n do
   begin
      P(i,0) := 0;
      for j := 1 to m do
         P(i,j) := P(i,j-1) + Q(G(i));
   end;
```

In this case, the computation of $P(i,j)$ depends on the value $P(i,j-1)$ as well as on the value of Q at $G(i)$. However, this index expression cannot be resolved until the program is executed and a value is provided for $G(i)$. Moreover, such a value may be different at each execution of the algorithm, as different inputs may be provided. (In this case we also need to ensure that G is well-defined and the value provided generates a valid index expression for Q.)

A data dependence of this type is called *dynamic* or *run-time* as, in general, its complete representation and analysis is not feasible before the algorithm is executed. The main problem with dynamic data dependencies is that, in general, they do not allow for the derivation of array designs statically. Therefore, in principle, dynamic problems and regular array synthesis are, at a simple level, incompatible.

However, a second and surprising result of the work presented in this book is that for restricted classes of dynamic problems the provision of a regular array solution in a systematic fashion and at compile time is still feasible. The approach we take will require some boundedness assumptions on the specification. For instance, if for the above segment of code we assume that the input values of $G(i)$ are contained in a finite range, we can define a regular array which will execute the algorithm for all the inputs in that range.

1.5   Outline of the Book

This book is organised as follows. In Chapter 2 we describe the design process in regular array synthesis, and introduce some basic definitions and properties. We also discuss at length technical issues related to the regularisation of non-uniform data dependencies as the basis for the development of the following chapters. We also give a brief survey of the most significant contributions to regular array synthesis which can be found in the literature.

In Chapter 3 we introduce integral problems and their regularisation. We explore the relation between integral and affine problems and discuss the advantages and limitations of the approach.

In Chapter 4 we formalise the step from static to dynamic problems, and introduce a subclass of dynamic problems for which we provide systematic regularisation techniques.

In Chapter 5 we present a number of case studies to illustrate the application of our techniques to well-known problems from the literature. For comparison, we also illustrate integral and dynamic problems which fall outside the scope of our techniques, and for which we provide *ad hoc* solutions.

In Chapter 6 we outline some possible developments of the work and draw a number of conclusions.

To make the book as self-contained as possible, in the Appendices we have included most of the mathematical notation, definitions and properties upon which our work is based.

Chapter 2

Regular Array Synthesis

Regular array synthesis is the process of transforming the functional description of an algorithm into its implementation as a regular array. We call this process the *design process* of the algorithm. Its basic steps, illustrated in Fig. 2.1, include[1]: *specification, analysis, regularisation, space-time mapping* and *implementation*.

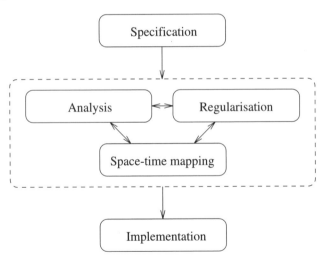

Fig. 2.1. Design process.

The specification of the algorithm denotes the first stage of the process, in which a functional description of the algorithm is provided as recurrence

---

[1]This is a simplified view of the design process. A more complete description is given in Section 2.4, including classes of transformations which are not considered in this book.

equations, or some equivalent such as single assignment code. The design process terminates with the implementation of the algorithm either as parallel code for a target machine or as specialised hardware. The remaining design steps (contained within the dotted box in the figure) constitute the core of the synthesis method, in which a regular spreading of the computations of the algorithm in space and time is obtained, so that the data dependence relations between computations are exposed and the maximal inherent parallelism of the algorithm can be exploited. A spatial distribution is obtained by representing the computations as the nodes of a graph. A temporal distribution follows from the partial order induced by the data dependence relations, and the assumption that a constant time cycle is associated with the execution of each computation. Regularity is enforced through systematic transformations of the specification.

In the figure, the arrows indicate the logical flow in the design process. That all arrows in the synthesis core are bi-directional follows from the fact that several iterations of the design steps may be necessary before the algorithm reaches a form which satisfies optimality, design and correctness requirements.

This chapter contains some preparatory work which is necessary to the development and understanding of the remainder of the book, and reviews some basic concepts of regular array synthesis which are of particular significance for our approach. The chapter is organised as follows. Section 2.1 contains some basic definitions relating to the specification, analysis and space-time mapping of regular parallel algorithms. In the discussion, we assume the existence of a generic index space for the representation of the computations of the algorithm, and develop the theory at an abstract level. In Section 2.2 we then specialise the theory to the case in which the computation space is embedded in a Euclidean (lattice) space. We hightlight the advantages of this embedding and give proofs of the relevant properties. Section 2.3 presents the basic issues underlying regularising transformations, laying the basis for the development of the following chapters. Section 2.4 contains a brief survey of the major contributions to regular array synthesis which can be found in the literature. Finally, Section 2.5 summarises the main issues discussed in this chapter.

2.1 Basic Design Steps

In this section, we introduce the basic concepts relative to the specification, analysis and space-time mapping of algorithms. We assume the existence of an index space with no particular structure (except for being a set) for representing the computations of an algorithm. We call it the *computation space* and denote it by $\mathcal{CS}$. We also assume the existence of a set Var of variable names, from which new variables can always be selected, and a set of data values Val, from which variables are assigned and which contains a special undefined value $\bot$. Variables are indexed by computation points. In particular, if V is a variable and $c \in \mathcal{CS}$, then $V(c)$ is called an *instance of* V at c. Also, a mapping from $\mathcal{CS}$ to itself is called an *index mapping*.

2.1.1 Algorithm Specification

In our context, an algorithm is specified as a system of recurrence equations, defining the computations and data dependencies of the algorithm. Each recurrence equation expresses the evaluation of a function, the *applied function*, on a set of computation points, the *computation domain*. The equations are recurrent in that the evaluation of a function may depend on other evaluations of the same function. Also, the equations constitute a system because, in general, the evaluation of a function depends on the evaluation of other functions of the system (so-called *simultaneous recursion*).

The result of evaluating a function on each point of a domain as prescribed by a recurrence equation, is recorded by an indexed variable, called the *result (variable)* of the recurrence equation. The indexing corresponds to the computation point. The definition of the applied function by a recurrence equation is obtained by specifying a number of indexed variables, *arguments*, to which the function is applied. The indexed variables provide a tabulation of the functions computed by the algorithm. The index expressions of the result and argument variables of a recurrence equation establish the *data dependence relations* existing between computations.

We distinguish two types of equations: recurrence equations, which define indexed variables in terms of functions applied to a number of argument indexed variables; and input equations, which define variables by the assignment of input values.

Definition 2.1.1 *[Recurrence and Input Equations]* A recurrence equation $\mathbf{E}$ (with m arguments) is defined as a 5-tuple $(D_{\mathbf{E}}, {}^{\bullet}\mathbf{E}, \mathbf{E}^{\bullet}, f_{\mathbf{E}}, \mathcal{IM}_{\mathbf{E}})$, where:

$D_{\mathbf{E}} \subseteq \mathcal{CS}$ is the computation domain of the equation; $^{\bullet}\mathbf{E} \in Var$ is the result of the equation; $\mathbf{E}^{\bullet} \in Var^m$ is the m-tuple of its arguments; $f_{\mathbf{E}} : Val^m \to Val$ is the applied function; and $\mathcal{IM}_{\mathbf{E}} \in [\mathcal{CS} \to \mathcal{CS}]^m$ is the m-tuple of its index mappings.

An input equation $\mathbf{E}$ is defined as a 3-tuple $(D_{\mathbf{E}}, {}^{\bullet}\mathbf{E}, f_{\mathbf{E}})$, where: $D_{\mathbf{E}} \subseteq \mathcal{CS}$ and $^{\bullet}\mathbf{E} \in Var$ are defined as above; and $f_{\mathbf{E}} : \mathcal{CS} \to Val$ is the applied function.   ■ 2.1.1

The applied function $f_{\mathbf{E}}$ has constant complexity and depends *strictly* on its arguments (see [Ka-et-al67]). Intuitively, $f_{\mathbf{E}}$ is strict on its arguments if $f_{\mathbf{E}}$ does not depend on computed values other than the value of its arguments. The notation $^{\bullet}\mathbf{E}$ and $\mathbf{E}^{\bullet}$ indicates the left- and right-hand sides of the equation $\mathbf{E}$. Given a recurrence equation $\mathbf{E}$, the *order of* $\mathbf{E}$ is equal to the number of its arguments, i.e., $|\mathbf{E}^{\bullet}| = m$ in the above definition. The order of an input equation is zero. Given an equation $\mathbf{E}$, we say that $\mathbf{E}$ defines variable $^{\bullet}\mathbf{E}$ on each point of the domain $D_{\mathbf{E}}$.

Before we proceed we should mention that in the literature there exists a variety of notations for recurrence equations (see, for instance, [Ka-et-al67, Rao85, QuVa89, Raj89, QuRo91]). Popular ones resemble the following:

$$c \in D : \quad U(c) = f(V_1(\mathcal{I}_1(c)), \ldots, V_m(\mathcal{I}_m(c)),$$

where D is the domain of the equation, U the result, f the applied function, and for all i, V_i are the arguments and $\mathcal{I}_i$ the index mappings. The correspondence between this notation and the one we adopt is straightforward, as illustrated in the example below. The reason why we have chosen an algebraic notation in this work is that it is particularly convenient to state the definitions and properties for the treatment of integral and dynamic recurrences.

Example 2.1.2   Consider the specification of the Fibonacci sequence given in Chapter 1. For convenience, we recall it here, both as imperative code:

```
F(0) := 1;
F(1) := 1;
for i := 2 to n do
    F(i) := F(i − 1) + F(i − 2);
```

and as the recurrence equation:

$$F(i) = \begin{cases} 1 & i = 0, 1 \\ F(i-1) + F(i-2) & i = 2, \ldots, n \end{cases}$$

In our notation, the specification can be expressed as:

$$\mathbf{E}_1 \;=\; (D_1, F, in_F)$$
$$\mathbf{E}_2 \;=\; (D_2, F, (F, F), +, (\mathcal{I}_1, \mathcal{I}_2))$$

where $D_1 = \{0, 1\}$, $D_2 = \{2, \ldots, n\}$, $in_F(i) = 1$, $+(a, b) = a + b$, $\mathcal{I}_1(i) = i - 1$ and $\mathcal{I}_2(i) = i - 2$,

which correspond to:

$$
\begin{aligned}
i \in D_1 : \quad & F(i) = \quad in_F(i) \\
i \in D_2 : \quad & F(i) = \quad +(F(\mathcal{I}_1(i)), F(\mathcal{I}_2(i)))
\end{aligned}
$$

∎ 2.1.2

An algorithm is specified as a system of recurrence and input equations. In such a system, each variable is assumed to be assigned exactly once at each computation point. This is enforced by assuming that equations in the system with the same result variable have disjoint domains. Note that the same property could be enforced in other ways, for instance by assuming that if the domains of such equations intersect, then the applied functions evaluate to the same value for each point of the intersection. Although the former condition is more restrictive, it is also easier to enforce and verify, hence it is the one which is usually adopted.

Definition 2.1.3 *[System of Equations]* A system of equations $\mathbf{S}$ is defined as a set of equations $\{\mathbf{E}_1, \ldots, \mathbf{E}_r\}$, where for all $\mathbf{E}_i, \mathbf{E}_j \in \mathbf{S}$, with $i \neq j$ and $i, j = 1, \ldots, r$, ${}^{\bullet}\mathbf{E}_i = {}^{\bullet}\mathbf{E}_j$ implies $D_{\mathbf{E}_i} \cap D_{\mathbf{E}_j} = \emptyset$. ∎ 2.1.3

Given a system of equations $\mathbf{S}$, we define its domain as the union of the domains of its equations, i.e., as the set $D_{\mathbf{S}} = \bigcup_{\mathbf{E} \in \mathbf{S}} D_{\mathbf{E}}$.

Example 2.1.4 *[System of Equations]* Consider the equations of Example 2.1.2 for the computation of the Fibonacci sequence. Equation $\mathbf{E}_1$ is an input equation, while $\mathbf{E}_2$ is a recurrence equation. Their domains are disjoint sets, so that $\mathbf{S} = \{\mathbf{E}_1, \mathbf{E}_2\}$ is a system of equations. ∎ 2.1.4

The set of variables of a recurrence equation $\mathbf{E}$ is defined as $Var_{\mathbf{E}} = \{{}^{\bullet}\mathbf{E}\} \cup \mathcal{S}(\mathbf{E}^{\bullet})$, where $\mathcal{S}(\mathbf{E}^{\bullet})$ is the support set of $\mathbf{E}^{\bullet}$. Informally, the support set of an m-tuple is the set obtained by combining the elements of the tuple. For instance, $\mathcal{S}((U, V, V, A)) = \{U, V, A\}$. (A formal definition of the support set $\mathcal{S}(\mathbf{u})$ of a tuple $\mathbf{u}$ is given in Appendix A). The set of variables of an input equation contains only the equation's result variable, that is $Var_{\mathbf{E}} = \{^{\bullet}\mathbf{E}\}$. The set of variables of a system of equation is the union of the set of variables of its equations, that is a system of equations $\mathbf{S}$ has set of variables $Var_{\mathbf{S}} = \bigcup_{\mathbf{E} \in \mathbf{S}} Var_{\mathbf{E}}$. Given a variable $V \in Var_{\mathbf{S}}$, we can identify the set of equations defining V, as $Def\!E_V = \{\mathbf{E} \in \mathbf{S} \mid {}^{\bullet}\mathbf{E} = V\}$, and the set of computation points on which V is defined, as $DefD_V = \bigcup_{\mathbf{E} \in Def\!E_V} D_{\mathbf{E}}$. In the following, we assume that an indexed variable has undefined value outside its definition domain, i.e., $V(c) = \perp$ for all $c \notin DefD_V$.

2.1.2   Analysis of the Data Dependencies

The process of extracting parallelism from an algorithm specification is based on the analysis of the *data dependencies* existing between its computations. To this end, several notions of dependence relation have been developed either among variables or their instances [Ka-et-al67, Rao85, SYKun88].

A basic dependence relation is the so-called *data dependence relation* or simply *data dependence*. This relation is defined between variable instances and is based on index mappings. A recurrence equation implicitly defines as many data dependencies as the order of the equation. Each of them involves the result of the equation and one of the arguments. If D is the domain of the equation, U its result, V one such argument and $\mathcal{I}$ the corresponding index mapping, a data dependence relation between U and V is established as follows. For each point c of the domain D, the computation of $U(c)$ depends on the value of V computed at the point $\mathcal{I}(c)$ of the computation space, hence the value $V(\mathcal{I}(c))$ has to be available for the computation of $U(c)$ to be possible.

Abstracting away from any particular equation, a data dependence can be represented as a 4-tuple $\mathcal{DD} = (D, U, V, \mathcal{I})$, where $D \subseteq \mathcal{CS}$, $U, V \in Var$ and $\mathcal{I} : \mathcal{CS} \to \mathcal{CS}$, which says that for all $c \in D$, $U(c)$ is data dependent on $V(\mathcal{I}(c))$.

Let $\mathbf{E}$ be a recurrence equation of order m. Then, the ith data dependence generated by $\mathbf{E}$ is $\mathcal{DD}_{\mathbf{E}_i} = (D_{\mathbf{E}}, \ {}^{\bullet}\mathbf{E}, pr_i(\mathbf{E}^{\bullet}), pr_i(\mathcal{IM}_{\mathbf{E}}))$, with $i \in 1, \ldots, m$. In this definition, pr_i represents the i_{th} projection applied to an m-tuple, that is pr_i returns the ith element of the m-tuple. For instance, $pr_3((U, W, V, S)) = V$. The set of data dependencies of a recurrence equation $\mathbf{E}$ is $\mathcal{DD}_{\mathbf{E}} = \bigcup_{i=1}^{m} \{\mathcal{DD}_{\mathbf{E}_i}\}$. As the order of an input equation is 0, its set of data dependencies is empty. Finally, the set of data dependencies of a system of equations $\mathbf{S}$ is $\mathcal{DD}_{\mathbf{S}} = \bigcup_{\mathbf{E} \in \mathbf{S}} \mathcal{DD}_{\mathbf{E}}$.

Example 2.1.5 *[Data Dependencies]* Consider the system **S** of Example 2.1.4. Its data dependencies are:

$$\begin{aligned} \mathcal{DD}_1 &= (D_2, F, F, \mathcal{I}_1) \\ \mathcal{DD}_2 &= (D_2, F, F, \mathcal{I}_2). \end{aligned}$$

∎ 2.1.5

Graphs are usually adopted for the representation and analysis of data dependencies. The following three types of graphs are commonly used in regular array synthesis[2]. A fourth type of graph, the so-called *signal flow graph*, which also plays a very important rôle in synthesis, will be introduced in Section 2.1.3 for the description of regular array designs.

Data Dependence Graph

A data dependence graph is the natural graph representation of a data dependence. Its nodes are computation points and its arcs relate computation points under the index mapping of the data dependence.

Definition 2.1.6 *[Data Dependence Graph]* Let $\mathcal{DD} = (D, U, V, \mathcal{I})$ be a data dependence. We define its data dependence graph $\mathcal{DDG}$ as the graph $(\mathcal{N}, \mathcal{A})$, where:

- $\mathcal{N} = D \cup \mathcal{I}(D)$; and

- $\mathcal{A} = \{(\mathcal{I}(c), c) \mid c \in D\}$.

∎ 2.1.6

The data dependence graph of an equation or a system of equations is obtained by combining the data dependence graphs of all their data dependencies. More precisely, it is obtained by taking the union of all the nodes and the disjoint union of all the arcs. This means that there may be more than one arc between each pair of nodes of the resulting graph. This is because distinct data dependencies may share the same domain. If there is a need to distinguish the arcs, then suitable labels should be used.

---

[2]Many variants of these graphs are known from the literature. According to Feautrier [Fea92a]: "Dependence graphs are used in every form of parallel programming [...]. There are nearly as many dependence graphs as there are workers in the field.".

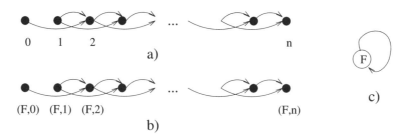

a)

b)

c)

Fig. 2.2. System **S**: a) data dependence graph; b) complete data dependence graph; c) reduced dependence graph.

Example 2.1.7 *[Data Dependence Graph]* Consider the system **S** of Example 2.1.4. Its data dependence graph, illustrated in Fig. 2.2 a), is $\mathcal{DDG} = (\mathcal{N}, \mathcal{A})$, where:

$$
\begin{aligned}
\mathcal{N} &= \{i, i-1, i-2 \mid i = 2, \ldots, n\} \\
\mathcal{A} &= \{(i-1, i), (i-2, i) \mid i = 2, \ldots, n\}.
\end{aligned}
$$

■ 2.1.7

Complete Data Dependence Graph

In a data dependence graph variable names are abstracted away: the nodes represent computation points, and distinct variable instances at the same computation point are not distinguished in the graph. When a distinction between variable instances is important, the following type of graph is used[3]:

Definition 2.1.8 *[Complete Data Dependence Graph]* Let $\mathcal{DD} = (D, U, V, \mathcal{I})$ be a data dependence. We define its complete data dependence graph $\mathcal{CDDG}$ as the graph $(\mathcal{N}, \mathcal{A})$, where:

- $\mathcal{N} = \{(U, c) \mid c \in D\} \cup \{(V, d) \mid d \in \mathcal{I}(D)\}$; and

- $\mathcal{A} = \{(V, \mathcal{I}(c)), (U, c)) \mid c \in D\}$.

■ 2.1.8

---

[3]The notation (U, c), where U is a variable name and c a computation point, is used to distinguish a node of the complete data dependence graph from $U(c)$, which represents the instance of U at c.

The complete data dependence graph of an equation or a system of equations is the union of the complete dependence graphs of their data dependencies, that is the union of both their nodes and arcs. Differently from the dependence graph, there can be at most one arc for each pair of nodes.

Example 2.1.9 *[Complete Data Dependence Graph]* Consider the system **S** of Example 2.1.4. Its complete data dependence graph, illustrated in Fig. 2.2 b), is $CDDG = (\mathcal{N}, \mathcal{A})$, where:

$$
\begin{aligned}
\mathcal{N} &= \{(F,i),(F,i-1),(F,i-2) \mid i=2,\dots,n\} \\
\mathcal{A} &= \{((F,i-1),(F,i)),((F,i-2),(F,i)) \mid i=2,\dots,n\}
\end{aligned}
$$

■ 2.1.9

Reduced Dependence Graph

Data dependence graph and complete data dependence graph allow a fine-grain analysis of the algorithm at the level of its basic computations. The reduced dependence graph, instead, allows a coarser-grain dependence analysis, in which the inter-dependencies between variables (hence, computed functions) are exposed.

Definition 2.1.10 *[Reduced Dependence Graph]* Let **S** be a system of equations. We define its reduced dependence graph $\mathcal{RDG}$ as the graph $(\mathcal{N}, \mathcal{A})$, where:

– $\mathcal{N} = Var_{\mathbf{S}}$; and

– $\mathcal{A} = \{(U,V) \mid \exists D, \mathcal{I} \text{ such that } (D,U,V,\mathcal{I}) \in \mathcal{DD}_{\mathbf{S}}\}$.

■ 2.1.10

Example 2.1.11 *[Reduced Dependence Graph]* Consider the system **S** of Example 2.1.4. Its reduced dependence graph, illustrated in Fig. 2.2 c), is $\mathcal{RDG} = (\mathcal{N}, \mathcal{A})$, where $\mathcal{N} = \{F\}$ and $\mathcal{A} = \{(F,F)\}$.   ■ 2.1.11

2.1.3   Space-Time Mapping

A (processor) array design for a specification is obtained by mapping its data dependence graph onto a graph describing the design. The mapping is realised by identifying a *timing function*, or *scheduling*, and an *allocation function*, or *placement*, of the computations.

A timing and allocation pair are usually selected on the basis of minimising some parameters, such as the number of execution steps of the algorithm, the number of processing elements of the network, or their connections and/or memory requirements. The selection of an optimal space-time mapping is of fundamental importance in the synthesis process, and has been investigated extensively (see, e.g., [Da-et-al91, ShFo92, DaRo94]). Whilst in this section we only recall some of the basic requirements and properties of a space-time mapping, in the following section we will discuss how optimal timing functions can be derived systematically for particular forms of recurrence equations embedded in Euclidean lattice spaces.

Timing Function

A timing function associates an execution time with each computation point of a system of equations. We only consider discrete schedulings, hence we may assume that time coincides with the integers, so that a timing function will be a partial mapping from CS to $\mathbf{Z}$. In particular, a timing function for a system of equations is a mapping from CS to $\mathbf{Z}$ which is defined (at least) at each computation point of the system. A valid timing function for a system of equations is a timing function which preserves the ordering of the computations imposed by their data dependencies[4]:

Definition 2.1.12 *[Valid Timing Function for a Data Dependence Graph]*
Let $\mathcal{DDG} = (\mathcal{N}, \mathcal{A})$ be a data dependence graph and t a timing function for $\mathcal{DDG}$. Then t is a valid timing function for $\mathcal{DDG}$ if and only if for all $(c, c') \in \mathcal{A}$, $t(c) < t(c')$.   ■ 2.1.12

Given a timing function t, an infinite family of timing functions can be derived from t by introducing integral delays[5]. In fact, if t is a valid timing function for a data dependence graph $\mathcal{DDG}$, then for all $c \in CS$, the mapping defined by $t_\beta(c) = t(c) + \beta$, with $\beta \in \mathbf{Z}$, is a valid timing function for $\mathcal{DDG}$. The proof rests on the observation that the ordering defined by a valid timing function among the nodes of a data dependence graph is invariant under the

---

[4]Feautrier, in [Fea92a], uses the name *causality* to indicate a condition similar to the one we use in the definition of valid timing function.

[5]Indeed, this is not the only operation which preserves timing functions. It is, however, the only operation we will consider in this work.

addition of constants. We will use this invariant to restrict ourselves to valid timing functions which are non-negative.

For implementation on finite array architectures, the following two properties of a timing function are required: that only finite sets of computations are scheduled for parallel execution; and that there is a starting time for the execution of the algorithm. For specifications with a finite computation set any valid timing function satisfies these requirements.

Definition 2.1.13 *[Finite and Bounded Timing Function]* Let $\mathcal{DDG} = (\mathcal{N}, \mathcal{A})$ be a data dependence graph and t a valid timing function for $\mathcal{DDG}$. Then:

- t is *finite* if and only if for all $\tau \in range_t(\mathcal{N})$, the set $\{c \in \mathcal{N} \mid t(c) = \tau\}$ is finite;

- t is *bounded (below)* if and only if there exists $\bar{\tau} \in range_t(\mathcal{N})$ such that for all $\tau \in range_t(\mathcal{N})$, $\bar{\tau} \leq \tau$.

■ 2.1.13

In this definition, $range_t(\mathcal{N})$ is the range of t over the set $\mathcal{N}$, that is the set of values $t(n)$ for each $n \in \mathcal{N}$.

From the previous discussion, it follows that, given a valid bounded timing function t with lower bound $\bar{\tau}$, it is always possible to derive a non-negative valid bounded timing function from t, for instance, by adding a delay equal to $\bar{\tau}$. Without loss of generality, in the following we will restrict ourselves to non-negative timing functions, unless otherwise specified.

The above definitions naturally extend from a data dependence graph $\mathcal{DDG}$ to, respectively, the data dependence $\mathcal{DD}$, equation $\mathbf{E}$ or system of equations $\mathbf{S}$, of which $\mathcal{DDG}$ is the data dependence graph. Hence, for instance, we may refer to a valid timing function of a system of equations, by which we mean a valid timing function of its data dependence graph. From hereon, in our discussion we will assume that all timing functions are valid.

Example 2.1.14 *[Timing Function]* Consider the data dependence graph in Fig. 2.3 a). A valid timing function is defined by: $t(i) = 0$, for $i = 1, \ldots, 8$; $t(i) = 1$, for $i = 9, \ldots, 12$; $t(i) = 2$, for $i = 13, 14$; and $t(15) = 3$.

■ 2.1.14

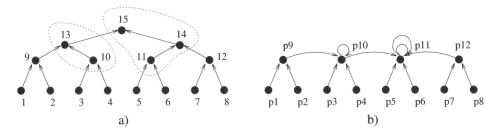

Fig. 2.3. a) Data dependence graph; b) signal flow graph.

Allocation Function

An allocation function determines the distribution of the computations of a system of equations among a set of processing elements.

We assume the existence of an index space, which we call the *processor space* and denote by $\mathcal{PS}$, as a convenient abstraction in which the placement of computations onto processing elements can be represented (in the same way the space $\mathcal{CS}$ provides a convenient abstraction for the representation of computations). No particular structure is assumed of $\mathcal{PS}$, other than being a set.

An allocation function can be represented as a partial mapping from $\mathcal{CS}$ to $\mathcal{PS}$. As for timing functions, an allocation function is required to be defined at least on the computation points of a system of equations.

We term a timing and an allocation functions as *compatible*[6] if and only if computations which are scheduled at the same time are not allocated to the same processing element, i.e., potentially parallel computations are not forced to be executed sequentially. This implies that the maximal parallelism of the specification is exploited, and no sequentiality is enforced which is not prescribed by the partial order defined by the data dependencies of the algorithm[7]. The property of compatibility may be expressed more formally as follows:

Definition 2.1.15 *[Compatibility]* Let $\mathcal{DDG} = (\mathcal{N}, \mathcal{A})$ be a data dependence graph and let t be a timing function and a an allocation function for

---

[6]Another term used in the literature is *conflict-free* [Fea92a].

[7]Note that with this notion of compatibility, design constraints related to the implementation of the algorithm, such as a limited number of processing elements, are not taken into consideration. See also the discussion on partitioning techniques in Sections 2.2.4 and 2.4.

$\mathcal{DDG}$. Then t and a are compatible if and only if for all $c, c' \in \mathcal{N}$, with $c \neq c'$, $t(c) = t(c')$ implies $a(c) \neq a(c')$. ∎ 2.1.15

Note that given a valid timing function t for a system of equation and the system's $\mathcal{DDG}$, any allocation function which is injective on the nodes of $\mathcal{DDG}$ is trivially compatible with t. However, as we will see in the following section, such allocation functions correspond to highly inefficient array designs.

Example 2.1.16 *[Allocation Function]* Consider the data dependence graph in Fig. 2.3 a) with valid timing function t as given in Example 2.1.14. An allocation function a compatible with t is defined as: $a(i) = p_i$, for $i = 1, \ldots, 9$; $a(10) = a(13) = p_{10}$; and $a(11) = a(14) = a(15) = p_{11}$; and $a(12) = p_{12}$, where p_i, for $i = 1, \ldots, 12$, denote distinct processors in $\mathcal{PS}$. In Fig. 2.3 a), dotted lines surround computation points which are allocated to the same processing element under a.

A non-compatible allocation function is, for instance, a' defined as: $a'(i) = p_i$, for $i = 1, \ldots, 8$; $a'(9) = a'(13) = p_9$, $a'(10) = a'(11) = p_{10}$; and $a'(12) = a'(14) = a'(15) = p_{11}$. Under a', computations 10 and 11 would contend the use of processor p_{10} at time $t = 2$.

A trivially compatible allocation function is a'' defined as $a''(i) = p_i$, for $i = 1, \ldots, 15$, with p_i distinct processors in $\mathcal{PS}$, for $i = 1, \ldots, 15$. ∎ 2.1.16

In the above presentation, we have assumed that a timing function is selected first, followed by the choice of a compatible allocation function. In this approach, the emphasis is on the optimality of the scheduling as a timing function which guarantees the minimum number of computation steps. Alternatively, an allocation function may be selected first [ClMo93]. In this case, the optimality of a placement can be expressed in terms of a number of design parameters, such as the locality of the connections of the network or the memory requirements for its processing elements.

Signal Flow Graph

A valid timing function for a system of equations together with a compatible allocation function determine an array design for a system of equations. Such a design is represented by a labelled graph known as a signal flow graph[8].

---

[8]Signal flow graphs are well-known design tools for digital signal processing applications. The type of signal flow graph used in this work is one particular type of signal flow graph, and other definitions may be found, for instance, in [CrRa83, SYKun88].

Its nodes represent processing elements, its edges communication links and their labels communication delays.

Definition 2.1.17 *[Signal Flow Graph under t and a]* Let **S** be a system of equations and $\mathcal{DDG} = (\mathcal{N}, \mathcal{A})$ its data dependence graph. Let t be a valid timing function for **S** and a a compatible allocation function. Define the signal flow graph $\mathcal{G}^{t,a}$ of **S** under t and a, as the labelled graph $(\mathcal{N}^{t,a}, \mathcal{A}^{t,a}, \ell)$ such that:

- $\mathcal{N}^{t,a} = \{a(c) \mid c \in \mathcal{N}\}$;

- $\mathcal{A}^{t,a} = \{(a(c), a(c')) \mid (c, c') \in \mathcal{A}\}$; and

- $\ell : \mathcal{A}^{t,a} \to \mathbf{N}$ such that $\forall (c, c') \in \mathcal{A}$, $\ell((a(c), a(c'))) = t(c') - t(c)$.

∎ 2.1.17

Note that as the dependence graph of a system, the signal flow graph may, and generally does, include more than one arc for each pair of nodes, and extra labels should be used if a distinction between the arcs is required.

Given a timing function for a system of equations, and the system's data dependence graph, an allocation function which is trivially compatible (i.e., an allocation which is just an injective mapping on the nodes of the graph – see Section 2.1.3) generates a highly inefficient array design as its processing elements are active exactly once.

Example 2.1.18 *[Signal Flow Graph]* Consider the data dependence graph in Fig. 2.3 a), the timing function t, given in Example 2.1.14, and the compatible allocation function a, given in Example 2.1.16. The corresponding signal flow graph is illustrated in Fig. 2.3 b). All the arcs have a unit label, which we have omitted in the figure. In this (and in the general) case, a associates more than one computation to some of the nodes of the signal flow graph, and the correct order of execution of such computations is determined by the timing function t. ∎ 2.1.18

Note that a signal flow graph provides an abstraction for the algorithm implementation. Such an implementation can be realised in a number of ways, including custom VLSI circuits whose topology matches that of the signal flow graph, or parallel code to be executed on a general purpose parallel machine.

Timing Function and Cyclic Data Dependence Graph

Definition 2.1.12 precludes the definition of a valid timing function for a data dependence graph containing cycles. Let $\mathcal{DDG}$ be a data dependence graph containing a path $n_1, \ldots, n_p$ such as $n_1 = n_p = n$, that is the path is a cycle. Let t be a timing function for $\mathcal{DDG}$. If we assume that t is valid then $t(n_1) < t(n_p)$, i.e., $t(n) < t(n)$. But this cannot be the case as t is a function of the integers. As a consequence only acyclic data dependence graphs admit valid timing functions.

Requiring that the data dependence graph is acyclic is too strong a restriction: there are many examples of recurrences with cyclic data dependence graphs which are (intuitively) computable and for which we would like to define a timing function. For instance:

Example 2.1.19 The system $\mathbf{S} = \{\mathbf{E}_1, \ldots, \mathbf{E}_5\}$ of equations below computes the sum $a + b$, where a and b are input values (at points 1 and n, respectively). The sum is computed at each point of the sub-domain D_2 of $\mathbf{S}$ (we have intentionally defined $\mathbf{S}$ so that its data dependence graph is cyclic, although simpler definitions no doubt exist). The equations are:

$$
\begin{aligned}
\mathbf{E}_1 &= (D_1, A, in_a) \\
\mathbf{E}_2 &= (D_2, A, A, id, \mathcal{I}_1) \\
\mathbf{E}_3 &= (D_3, B, in_b) \\
\mathbf{E}_4 &= (D_2, B, B, id, \mathcal{I}_2) \\
\mathbf{E}_5 &= (D_2, F, (A, B), +, (\mathcal{I}, \mathcal{I})),
\end{aligned}
$$

with: domains $D_1 = \{1\}$, $D_2 = \{2, \ldots, n-1\}$, $D_3 = \{n\}$; index mappings $\mathcal{I}(i) = i$, $\mathcal{I}_1(i) = i - 1$, $\mathcal{I}_2(i) = i + 1$; computed functions $id(a) = a$, $in_a(i) = a$, $in_b(i) = b$, with $a, b \in \mathbf{R}$, and $+(a, b) = a + b$. The addition $a + b$ at each point of D_2 is realised by the computation of variable F. Variables A and B transfer, respectively, a and b among the computation points. The data dependence graph and complete data dependence graph of $\mathbf{S}$ are illustrated in Fig. 2.4 a) and b), respectively. Note that while the data dependence graph of $\mathbf{S}$ is cyclic, the complete data dependence graph of $\mathbf{S}$ is not.   ■ 2.1.19

The reason why cycles may appear in a data dependence graph even when the complete data dependence graph is acyclic is that a data dependence graph abstracts away from variable names, i.e., does not distinguish

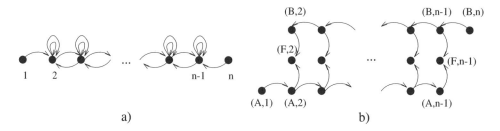

a)   b)

Fig. 2.4. System **S**: a) data dependence graph; b) complete data dependence graph.

between variables, so that variable instances with different names at the same computation point are mapped onto the same node of the graph. Such cycles can be eliminated if the specification is modified, for instance, to redistribute its computations in the space, so that a distinct computation point is associated with each node of the acyclic complete data dependence graph.

In the following, without loss of generality and unless otherwise indicated, we will always assume that data dependence graphs are acyclic.

2.2   Euclidean Synthesis

Although in the previous discussion we made no assumptions on the nature and properties of the computation space $\mathcal{CS}$ and processor space $\mathcal{PS}$, regular array synthesis techniques have developed mainly for computation and processor spaces represented by multi-dimensional Euclidean spaces. This geometric embedding of regular array synthesis has the following important advantages. From a theoretical point of view, linear algebra and affine geometry provide for rich mathematical models. From an applicative point of view, their constructive and algorithmic properties may be exploited for the development of automatic support.

The central idea of Euclidean regular array synthesis is the embedding of the data dependence graph of a specification into convex regions of the space. Of such convex regions only lattice points are considered. Therefore, the actual representation of the graph is in a lattice space. However, properties of the embedding linear and affine spaces can be exploited, such as polyhedral convexity, which allows for a finite representation of the computation domains, and linear and affine transformations, which, by preserving polyhedral convexity, allow for a finite manipulation of the domains.

The choice of linear and affine (lattice) spaces for the representation of algorithm specifications is not arbitrary, but comes from the realisation that multi-dimensional lattice spaces offer a natural interpretation of the index spaces defined by sets of nested for-loops, which, in turn, are widely used in the coding of large classes of numerical algorithms [Lam74]. Moreover, for those algorithms, loop bounds are often given as affine expressions of the indices, which can be interpreted as the definition of the boundaries of polyhedral convex sets in the embedding Euclidean space.

Finally, another major benefit of this representation is the possibility of defining systematic techniques for generating optimal space-time mappings, by relating the problem specification to the formulation of (integer) linear programming problems, for which computationally effective solution methods are well-developed [Sch86].

In the remainder of this section we will discuss the basic concepts of Euclidean synthesis and the specialisation of regular array synthesis concepts arising from this embedding strategy. From now on, computation and processor spaces are assumed to be multi-dimensional Euclidean spaces and computation domains to be (convex) polyhedra.

2.2.1 Representation and Mechanisation

For a specification to be formally represented and manipulated, it has to be expressible as a term according to some syntactic rules. Euclidean synthesis techniques account for this requirement. In particular computation domains, index mappings, and timing and allocation functions admit term representations, through linear and affine expressions.

As computation domains are assumed to be polyhedral convex sets, by definition, they admit a finite representation either as the intersections of a finite set of half-spaces or as the convex closure of a finite number of points and directions (see also Appendix C).

Index mappings, in their most general form, are assumed to be affine transformations. Hence they can be represented as a combination of a linear transformation (as a matrix-vector product) and a translation vector. A similar form is provided for timing and allocation functions, which are assumed to be affine and linear transformations, respectively. These forms and related properties are discussed in the remainder of this section.

2.2.2   Normalisation of Index Expressions

In Euclidean synthesis the computation domain is assumed to be a multi-dimensional lattice space. Let $\mathbf{Z}^n$ denote such a space for some natural $n \in \mathbf{N}$. Then an index mapping in $\mathbf{Z}^n$ is a mapping $\mathcal{I}$ from $\mathbf{Z}^n$ to $\mathbf{Z}^n$.

In general, however, the initial specification of an algorithm may contain a more general form of index mapping as a mapping $\mathcal{I} : \mathbf{Z}^n \to \mathbf{Z}^l$, where $l \leq n$. Such mappings need to be *normalised* [QuVa89] before synthesis techniques can be applied. For example, let us consider the specification of the matrix product of Chapter 1 (recalled here, for convenience):

```
for i := 1 to n do
for j := 1 to n do
  begin
    C(i, j, 0) := 0;
    for k := 1 to n do
      C(i, j, k) := C(i, j, k − 1) + A(i, k) * B(k, j);
  end;
```

Its computation space can be assumed to be a 3-dimensional space in which each value $C(i, j, k)$ is computed for all i, j, k from 1 to n. Note that the number of dimensions of the space corresponds to the number of nested for-loops. Variables A and B are indexed by two indices only and need to be normalised in $\mathbf{Z}^3$.

Normalisation usually consists of "padding" the index expressions of variables which are not fully indexed, with extra indices up to the number n of dimensions of the space. Geometrically, this corresponds to positioning the computations of such variables in particular sub-spaces of the n-dimensional space. In principle, any arbitrary choice of the extra indices can be made. However, because of the relation between index expressions and data dependencies, this choice has an impact on the regularity of the specification. When the variables to be fully indexed correspond to input or output data of the algorithm (as in our example), it is common practice to position their evaluation on some boundary of the computation domains. This, in general, results in array designs where input and output operations are confined to boundary processing elements.

In our examples, we could pad the index expressions of A and B with extra null entries and obtain:

```
for i := 1 to n do
```

```
for j := 1 to n do
   begin
      C(i, j, 0) := 0;
      for k := 1 to n do
         C(i, j, k) := C(i, j, k − 1) + A(i, k, 0) ∗ B(k, j, 0);
   end;
```

We can consider the overall computation domain to be the hypercube $D = \{(i, j, k) \mid 1 \leq i, j \leq n, 0 \leq k \leq n\}$, which corresponds to the evaluation of all index expressions in the nested for-loops, for i, j and k ranging between their bounds. Then the evaluation of A and B occurs on the boundary $D' = \{(i, j, k) \in D \mid k = 0\}$.

This type of normalisation is generally applicable [QuVa89] and, in the following, we always assume that all variables are normalised.

2.2.3 Uniform and Affine Data Dependencies

An affine index mapping is an index mapping in $\mathbf{Z}^n$ which defines an affine transformation. An affine index mapping has the following form:

Definition 2.2.1 *[Affine Index Mapping]* Let $\mathcal{I}$ be an index mapping. $\mathcal{I}$ is affine if for all $z \in \mathbf{Z}^n$, $\mathcal{I}(z) = A \cdot z + b$, with $A \in \mathbf{Z}^{n \times n}$ and $b \in \mathbf{Z}^n$. In addition, $\mathcal{I}$ is uniform if $A = \mathbf{I}_n$.   ■ 2.2.1

Together with index mappings, so-called *dependence mappings* are commonly used in the analysis of the data dependencies. Dependence mappings are specific to Euclidean synthesis in that their definition relies upon the existence of arithmetic operations in the computation space. A dependence mapping is derived from an index mapping as follows:

Definition 2.2.2 *[Affine Dependence Mapping]* Let $\mathcal{I}$ be an affine index mapping. Its dependence mapping $\Theta_{\mathcal{I}}$ is the mapping defined as $\Theta_{\mathcal{I}}(z) = z - \mathcal{I}(z)$, for all $z \in \mathbf{Z}^n$.   ■ 2.2.2

For all z, $\Theta_{\mathcal{I}}(z)$ is a vector, called the *data dependence vector* determined by $\mathcal{I}$ at z. Note that as $\mathcal{I}$ is affine, $\Theta_{\mathcal{I}}$ is also affine. Affine index mappings are used in the definition of affine data dependencies:

Definition 2.2.3 *[Affine Data Dependence]* Let $\mathcal{DD} = (D, U, V, \mathcal{I})$ be a data dependence. $\mathcal{DD}$ is affine if $\mathcal{I}$ is an affine index mapping.   ∎ 2.2.3

Given an affine data dependence $\mathcal{DD} = (D, U, V, \mathcal{I})$, with $\Theta_{\mathcal{I}}$ the dependence mapping defined by $\mathcal{I}$, the image of D under $\Theta_{\mathcal{I}}$ is called the *dependence domain* of $\mathcal{DD}$:

Definition 2.2.4 *[Dependence Domain]* Let $\mathcal{DD} = (D, U, V, \mathcal{I})$ be an affine data dependence and $\Theta_{\mathcal{I}}$ the dependence mapping defined by $\mathcal{I}$. The dependence domain $\Omega_{\mathcal{I}}$ of $\mathcal{DD}$ is $\Omega_{\mathcal{I}} = \Theta_{\mathcal{I}}(D)$.   ∎ 2.2.4

As D is a convex polyhedron and $\Theta_{\mathcal{I}}$ is affine, $\Omega_{\mathcal{I}}$ is also a convex polyhedron (see Appendix C).

An important sub-class of affine data dependencies is represented by uniform data dependencies. A uniform data dependence is characterised by a constant vector replicated for each point of the domain. Because of their regularity, uniform data dependencies generate regular array designs under a space-time mapping. Uniformity depends both on the index mapping and the domain of the data dependence:

Definition 2.2.5 *[Uniform Data Dependence]* Let $\mathcal{DD} = (D, U, V, \mathcal{I})$ be an affine data dependence. $\mathcal{DD}$ is uniform if and only if there exists an integer vector $c \in \mathbf{Z}^n$ such that, for all $z \in D$, $\Theta_{\mathcal{I}}(z) = c$.   ∎ 2.2.5

From this definition it follows that the dependence domain of a uniform data dependence reduces to a singleton set.

Example 2.2.6 *[Uniform and Non-Uniform Data Dependencies]* Consider the index mapping

$$\mathcal{I} \begin{pmatrix} i \\ j \\ k \end{pmatrix} = \begin{pmatrix} k \\ j-1 \\ k \end{pmatrix} = \begin{bmatrix} 0 & 0 & 1 \\ 0 & 1 & 0 \\ 0 & 0 & 1 \end{bmatrix} \begin{pmatrix} i \\ j \\ k \end{pmatrix} + \begin{pmatrix} 0 \\ -1 \\ 0 \end{pmatrix},$$

in $\mathbf{Z}^3$, and its dependence mapping $\Theta_{\mathcal{I}}(i, j, k) = (i - k, 1, 0)$. Let $\mathcal{DD} = (D, U, V, \mathcal{I})$ and $\mathcal{DD}' = (D', U, V, \mathcal{I})$ be data dependencies with domains, respectively:

$$\begin{aligned} D &= \{(i, j, k) \mid 1 \le i, j \le m, 1 \le k \le i\} \\ D' &= \{(i, j, k) \mid 1 \le i, j \le m, k = i\}, \end{aligned}$$

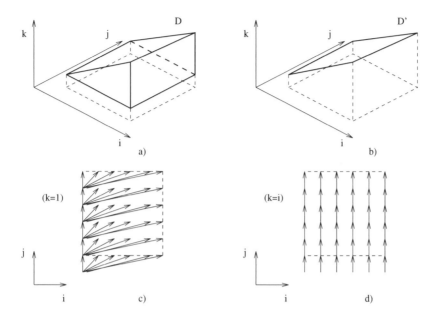

Fig. 2.5. Domains: a) D; and b) D'. Data dependence graph of: c) $\mathcal{DD}$ (section only); and d) $\mathcal{DD}'$.

where m is a value in $\mathbf{N}$. Domains D and D' are illustrated in Fig. 2.5 a) and b), respectively. A section (for $k = 1$) of the data dependence graph relative to $\mathcal{DD}$ is given in Fig. 2.5 c), while the data dependence graph of $\mathcal{DD}'$ is illustrated in part d) of the figure. Note that $\mathcal{DD}$ is not uniform. For instance, by assuming $m \geq 3$ and considering the points $(2, 2, 1)$ and $(3, 2, 1)$ of D, then $\Theta(2, 2, 1) = (1, 1, 0) \neq \Theta(3, 2, 1) = (2, 1, 0)$. However, $\mathcal{DD}'$ is uniform as for all $(i, j, k) \in D'$, $\Theta_{\mathcal{I}}(i, j, k) = (0, 1, 0)$.   ■ 2.2.6

The following proposition provides a necessary and sufficient condition for the uniformity of a data dependence:

Proposition 2.2.7 *[Condition for Uniformity]* Let $\mathcal{DD} = (D, U, V, \mathcal{I})$ be an affine data dependence, with $\mathcal{I}(z) = A \cdot z + b$, for all $z \in D$. Then $\mathcal{DD}$ is uniform if and only if $lin(D) \subseteq null(\mathbf{I}_n - A)$.

PROOF: By definition, the affine closure of D is:

$$aff(D) = \{\sum_i a_i z_i \mid z_i \in D, \sum_i a_i = 1\}$$

and given $d \in D$, the direction of D is:

$$lin(D) = aff(D) - d = \{z - d \mid z \in D\}.$$

From the above definitions, it follows that $\forall z \in D$, $z - d \in lin(D)$.

Let us prove that $lin(D) \subseteq null(\mathbf{I}_n - A)$ implies that there exists $c \in \mathbf{Z}^n$ such that $\forall z \in D$, $\Theta_{\mathcal{I}}(z) = c$. For all $z \in D$,

$$
\begin{aligned}
\Theta_{\mathcal{I}}(z) &= (\mathbf{I}_n - A) \cdot z - b = (\mathbf{I}_n - A) \cdot (z - d + d) - b \\
&= (\mathbf{I}_n - A) \cdot (z - d) + (\mathbf{I}_n - A) \cdot d - b.
\end{aligned}
$$

As $z - d \in lin(D)$ and, by assumption, $lin(D) \subseteq null(\mathbf{I}_n - A)$, then $(\mathbf{I}_n - A) \cdot (z - d) = 0$ and the expression above reduces to

$$\Theta_{\mathcal{I}}(z) = (\mathbf{I}_n - A) \cdot d - b,$$

where $(\mathbf{I}_n - A) \cdot d - b$ is a constant vector in $\mathbf{Z}^n$, independent from z. In particular, when $A = \mathbf{I}_n$, then $\Theta_{\mathcal{I}}(z) = -b$.

We now prove that the existence of $c \in \mathbf{Z}^n$ such that $\forall z \in D$, $\Theta_{\mathcal{I}}(z) = c$ implies $lin(D) \subseteq null(\mathbf{I}_n - A)$. As $\forall z \in D$, $\Theta_{\mathcal{I}}(z) = c$ then $\forall z \in D$, $(\mathbf{I}_n - A) \cdot z = b + c$, and, in particular, $(\mathbf{I}_n - A) \cdot d = b + c$. The (matrix) expression:

$$(\mathbf{I}_n - A) \cdot z = b + c$$

defines a system of non-homogeneous linear equations whose so-lution space $null(\mathbf{I}_n - A) + (b + c)$ contains D. Also $aff(D)$ is contained in this space, as any affine combination of elements of D is a solution to the system. In fact, let $z_i \in D$, for $i = 1\ldots,m$, and let $\sum_i a_i z_i$ be an affine combination of the points, with $\sum_i a_i = 1$. Then:

$$
\begin{aligned}
(\mathbf{I}_n - A) \cdot \left(\sum_i a_i z_i\right) &= \sum_i a_i (\mathbf{I}_n - A) \cdot z_i \\
&= \sum_i a_i (b + c) = b + c,
\end{aligned}
$$

i.e., the affine combination is a solution to the system. By definition of $lin(D)$, for all $z \in lin(D)$, there exist $z' \in aff(D)$ and $d \in D$ such that $z = z' - d$. Then:

$$(\mathbf{I}_n - A) \cdot z = (\mathbf{I}_n - A) \cdot (z' + d) = 0$$

i.e., $z \in null(\mathbf{I}_n - A)$. It follows that $lin(D) \subseteq null(\mathbf{I}_n - A)$.

■ 2.2.7

Corollary 2.2.8   Let $\mathcal{DD} = (D, U, V, \mathcal{I})$ be an affine data dependence. If $\mathcal{I}$ is uniform then $\mathcal{DD}$ is uniform.   ■ 2.2.8

Corollary 2.2.9   Let $\mathcal{DD} = (D, U, V, \mathcal{I})$ be an affine data dependence, with a non-uniform index mapping $\mathcal{I}$. If $\mathcal{DD}$ is uniform then $\mathcal{I}$ can be replaced by a uniform index mapping $\mathcal{I}'$.

PROOF: $\mathcal{I}'$ is the index mapping $\mathcal{I}'(z) = z - c$, where c is the constant vector of Proposition 2.2.7.   ■ 2.2.9

Corollary 2.2.9 states that if $\mathcal{DD}$ is uniform data dependence then its index mapping $\mathcal{I}$ can always be expressed as a uniform index mapping. This is illustrated in the following example:

Example 2.2.10   In Example 2.2.6, $lin(D) = \mathbf{Z}^3$, $lin(D') = \langle (1,0,1), (0,1,$
$0) \rangle$, $A = \begin{bmatrix} 0 & 0 & 1 \\ 0 & 1 & 0 \\ 0 & 0 & 1 \end{bmatrix}$, and $null(\mathbf{I}_3 - A) = \langle (1,0,1), (0,1,0) \rangle$. As $lin(D) \not\subseteq$
$null(\mathbf{I}_3 - A)$, $\mathcal{DD}$ is not uniform. As $lin(D') \subseteq null(\mathbf{I}_3 - A)$ (in particular they are equal), $\mathcal{DD}'$ is uniform. Note that, in $\mathcal{DD}'$, $\mathcal{I}$ can be replaced by $\mathcal{I}'(i, j, k) = (i, j, k) + (0, -1, 0)$.   ■ 2.2.10

2.2.4   Affine Space-Time Mapping

An affine timing function is a timing function which is also an affine transformation:

Definition 2.2.11 *[Affine Timing Function]* Let $t : \mathbf{Z}^n \to \mathbf{Z}$ be a timing function. t is affine if and only if, for all $z \in \mathbf{Z}^n$, $t(z) = \lambda \cdot z + \mu$, where λ is a non-null vector in $\mathbf{Z}^n$ and $\mu \in \mathbf{Z}$.   ■ 2.2.11

The vector λ may be regarded as the normal vector to a family of parallel hyperplanes which it defines. There exists one such hyperplane for each (instant) $i \in \mathbf{Z}$, called an *isochronous* or *equitemporal* hyperplane. Intuitively, all the points on such a hyperplane are assigned by t the same instant of time, and so should be computed on different processors.

A linear allocation function is an allocation function which is defined as a linear transformation from the computation space to the processor space. The processor space is usually assumed to have dimensionality no greater

than the computation space, as, in general, in order to increase the efficiency of the array, several (sequential) computations are allocated to the
same processor. Hence, in general, the image of an allocation function has a
lower dimensionality than its application domain. Computationally, a mapping from $\mathbf{Z}^n$ onto a space of lower dimension can be conveniently achieved
through a linear projection.

Definition 2.2.12 *[Linear Allocation Function]* Let $a : \mathbf{Z}^n \to \mathbf{Z}^l$ be an
allocation function. a is linear if and only if, for all $z \in \mathbf{Z}^n$, $a(z) = \sigma \cdot z$,
where σ is a non-null matrix in $\mathbf{Z}^{l \times n}$ and $l \leq n$.   ■ 2.2.12

The simplest type of linear allocation functions are projections from $\mathbf{Z}^n$
to $\mathbf{Z}^{n-1}$. These projections can be expressed by corresponding projection
vectors:

Definition 2.2.13 *[Projection Vector]* Let $a : \mathbf{Z}^n \to \mathbf{Z}^{n-1}$ be a linear
allocation function, with $a(z) = \sigma \cdot z$ and $\sigma \in \mathbf{Z}^{(n-1) \times n}$, for all $z \in \mathbf{Z}^n$. A
projection vector for a is a non-null vector $u \in null(\sigma)$.   ■ 2.2.13

For affine timing and allocation functions t and a, respectively, their
compatibility can be verified by considering the dot product $\lambda \cdot u$ as stated
in the following proposition:

Proposition 2.2.14 *[Compatibility]* Let t be an affine timing function such
that $t(z) = \lambda \cdot z + \mu$, and a a linear allocation function with projection vector
u. Then t and a are compatible if and only if $\lambda \cdot u \neq 0$.

PROOF: We first prove that if t and a are compatible then $\lambda \cdot u \neq 0$. Assume
$\lambda \cdot u = 0$. Then $u \in null(\lambda)$. Therefore there exist $z, z' \in [\lambda : c]$, for
some $c \in \mathbf{Z}$, such that $z' = z + mu$, with $m \neq 0$, and $t(z) = t(z')$.
Then:

$$a(z') = \sigma \cdot z' = \sigma \cdot (z + mu) =$$
$$= \quad \sigma \cdot z + m\sigma \cdot u = \sigma \cdot z = a(z),$$

Hence, t and a are not compatible.

We now prove that if $\lambda \cdot u \neq 0$ then t and a are compatible. Assume
t and a non compatible. Then there exist z, z' such that $z \neq z'$,
$t(z) = t(z')$ and $a(z) = a(z')$. However, if $a(z) = a(z')$, then there

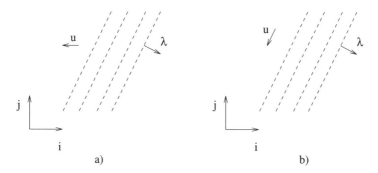

Fig. 2.6. a) Compatible and b) non-compatible affine scheduling and linear allocation.

exists $m \neq 0$ such that $z' = z + mu$, as u is the projection vector. Therefore:

$$t(z') = \lambda \cdot z' + \mu = \lambda \cdot (z + mu) + \mu =$$
$$= \lambda \cdot z + m\lambda \cdot u + \mu = t(z) + m\lambda \cdot u,$$

and $t(z') = t(z)$ only if $\lambda \cdot u = 0$. ∎ 2.2.14

That the dot product of λ and u is not zero, implies that points on an equitemporal hyperplane defined by λ are not projected onto the same processor, hence their computations can proceed in parallel. An intuitive picture of this property is given in Fig. 2.6 for λ and u in $\mathbf{Z}^2$.

An affine timing function t, defined as $t(z) = \lambda \cdot z + \mu$, together with a compatible linear allocation function a, defined as $a(z) = \sigma \cdot z$, can be regarded as a space-time mapping from $\mathbf{Z}^n$ to itself[9]. This mapping is usually represented in its matrix form as:

$$[t, a](z) = \begin{bmatrix} \lambda \\ \sigma \end{bmatrix}.$$

An affine space-time mapping $[t, a]$ defines a signal flow graph according to Definition 2.1.17. If $\mathbf{S}$ is a system of equations with dependence domain

---

[9]More general types of affine space-time mappings have been considered in the literature for partitioning techniques (see the discussion in Section 2.4). In such cases, both scheduling and allocation functions are seen as multi-dimensional mappings, with the dimensions of their ranges summing up to the dimension of the computation space. As we do not address partitioning in this work, and for ease of presentation, we do not consider this type of mapping formally.

Ω, the arcs of the signal flow graph under $[t, a]$ are the images of the vectors in Ω. In particular, each data dependence vector d in Ω corresponds to the arc $\sigma \cdot d$ of the signal flow graph, with label $\lambda \cdot d$.

2.2.5   Linear Optimisation and Affine Scheduling

One of the main advantages of Euclidean synthesis is the possibility of defining semi-automatic techniques for the scheduling of specifications containing affine data dependencies. Such techniques are based on the formulation of an integer linear programming problem, whose solution is an optimal affine timing function for the specification. These techniques were first introduced by Quinton in [Qui84] for uniform problems, and subsequently extended to affine problems in [RaFu87, RaFu89, QuVa89].

If $\mathcal{DD}$ is an affine data dependence, the optimisation problem which defines an optimal affine timing function for $\mathcal{DD}$ is defined both on the computation domain and the dependence domain of $\mathcal{DD}$. The fundamental result is expressed by theorem 2.2.20 that you will find towards the end of this section. Before presenting the theorem, however, we need a little more work, which is contained in the following two lemmas and their corollaries.

Lemma 2.2.15 *[Valid Affine Timing Function]* Let $\mathcal{DD} = (D, U, V, \mathcal{I})$ be an affine data dependence, $\Theta_{\mathcal{I}}$ its dependence mapping and $\Omega_{\mathcal{I}} = range_{\Theta_{\mathcal{I}}}$ (D). Let t be an affine timing function, such that for all $z \in \mathbf{Z}^n$, $t(z) = \lambda \cdot z + \mu$, with $\lambda \in \mathbf{Z}^n$ and with $\mu \in \mathbf{Z}$. t is a valid timing function for $\mathcal{DD}$ if and only if:

   i) for all $v \in vert(\Omega_{\mathcal{I}})$, $\lambda \cdot v > 0$;

   ii) for all $r \in ray(\Omega_{\mathcal{I}})$, $\lambda \cdot r \geq 0$.

PROOF: According to Definition 2.1.12, t is a valid timing function for $\mathcal{DD}$ if and only if for all $z \in D$, $t(z) > t(\mathcal{I}(z))$. By expanding t to its affine form we obtain:

$$\lambda \cdot z + \mu > \lambda \cdot \mathcal{I}(z) + \mu$$
$$\lambda \cdot (z - \mathcal{I}(z)) > 0$$
$$\lambda \cdot \Theta_{\mathcal{I}}(z) > 0$$

i.e., t is a valid timing function for $\mathcal{DD}$ if and only if for all $z \in D$, $\lambda \cdot \Theta_{\mathcal{I}}(z) > 0$.

First we prove that conditions i) and ii) implies $\lambda \cdot \Theta_\mathcal{I}(z) > 0$, for all $z \in D$. As $\Omega_\mathcal{I}$ is a convex polyhedron, it is finitely generated by a set of points and directions in $\mathbf{Z}^n$. Let $vert(\Omega_\mathcal{I}) = \{v_1, \ldots, v_m\}$ and $ray(\Omega_\mathcal{I}) = \{r_1, \ldots, r_p\}$. Then, for all $z \in D$, $\Theta_\mathcal{I}(z)$ can be expressed as:

$$\Theta_\mathcal{I}(z) = \sum_i a_i v_i + \sum_j b_j r_j,$$

with $a_i, b_j \geq 0$ and $\sum_i a_i = 1$. As $\Theta_\mathcal{I}(z) \neq 0$ (otherwise the data dependence graph is cyclic), from i), ii) and the conditions on a_i and b_j it follows that:

$$\lambda \cdot \Theta_\mathcal{I}(z) = \sum_i a_i \lambda \cdot v_i + \sum_j b_j \lambda \cdot r_j > 0.$$

We now prove that $\lambda \cdot \Theta_\mathcal{I}(z) > 0$, for all $z \in D$, implies conditions i) and ii). Condition i) is straightforward. As any vertex v of $\Omega_\mathcal{I}$ is in particular an element of the set, there exists $z \in D$ such that $v = \Theta_\mathcal{I}(z)$. Therefore $\lambda \cdot v > 0$. Condition ii) can be proved by noticing that if r is an infinite direction of $\Omega_\mathcal{I}$ then it is the image of an infinite direction r' of D under the linear part $\mathcal{L}_{\Theta_\mathcal{I}}$ of $\Theta_\mathcal{I}$. Suppose that for all z, $\Theta_\mathcal{I}(z) = \mathcal{L}_{\Theta_\mathcal{I}}(z) + c$, for some $c \in \mathbf{Z}^n$. Assume that $\lambda \cdot r < 0$ and consider $z + ar' \in D$. Then:

$$\begin{aligned}
\lambda \cdot \Theta_\mathcal{I}(z + ar') = & \\
= & \quad \lambda \cdot (\mathcal{L}_{\Theta_\mathcal{I}}(z + ar') + c) = \\
= & \quad \lambda \cdot (\mathcal{L}_{\Theta_\mathcal{I}}(z) + a\mathcal{L}_{\Theta_\mathcal{I}}(r') + c) = \\
= & \quad \lambda \cdot (\Theta_\mathcal{I}(z) + ar) = \\
= & \quad \lambda \cdot \Theta_\mathcal{I}(z) + a\lambda \cdot r,
\end{aligned}$$

which is negative for a sufficiently large, as $\lambda \cdot \Theta_\mathcal{I}(z) > 0$. As this violates the validity of t, then the assumption $\lambda \cdot r < 0$ is false.

∎ 2.2.15

Corollary 2.2.16   Let $\mathcal{DD} = (D, U, V, \mathcal{I})$ be a uniform data dependence, with index mapping $\mathcal{I}(z) = z + b$, for all $z \in \mathbf{Z}^n$, and $b \in \mathbf{Z}^n$. Let t be an affine timing function, such that for all $z \in \mathbf{Z}^n$, $t(z) = \lambda \cdot z + \mu$, with $\lambda \in \mathbf{Z}^n$ and $\mu \in \mathbf{Z}$. t is a valid timing function for $\mathcal{DD}$ if and only if $\lambda \cdot b < 0$.

∎ 2.2.16

Lemma 2.2.17 *[Finite and Bounded Affine Timing Function]* Let $\mathcal{DD} = (D, U, V, \mathcal{I})$ be an affine data dependence and t a valid affine timing function for $\mathcal{DD}$. Then:

i) t is finite if and only if, for all $r \in ray(D)$, $\lambda \cdot r \neq 0$;

ii) t is bounded (below) if and only if, for all $r \in ray(D)$, $\lambda \cdot r \geq 0$.

PROOF: i) As D is a convex polyhedral set, D contains unbounded sets of the form $\{z + ar | a \geq 0\}$, for $z \in D$ and $r \in ray(D)$. Therefore, t is not finite if and only if, for any such set, $t(z) = t(z + ar)$ for all $a > 0$. However, by expanding the definition of t,

$$\begin{aligned} t(z) &= \lambda \cdot z + \mu \\ t(z + ar) &= \lambda \cdot (z + ar) + \mu = \lambda \cdot z + \mu + a\lambda \cdot r \end{aligned}$$

Therefore, $t(z) = t(z + ar)$ for all $a > 0$ if and only if $\lambda \cdot r = 0$.

ii) t is not bounded below if and only if, given a set $\{z + ar | a \geq 0\}$, for $z \in D$ and $r \in ray(D)$, there exists an infinitely decreasing chain $t(z) > t(z + r) > \ldots > t(z + ar) > \ldots$ Such a chain exists if and only if $\lambda \cdot r < 0$. In fact, for all $a \geq 0$,

$$t(z + ar) > t(z + (a + 1)r)$$

if and only if

$$\begin{aligned} \lambda \cdot (z + ar) + \mu &> \lambda \cdot (z + (a + 1)r) + \mu \\ \lambda \cdot z + a\lambda \cdot r + \mu &> \lambda \cdot z + (a + 1)\lambda \cdot r + \mu \\ a\lambda \cdot r &> (a + 1)\lambda \cdot r \end{aligned}$$

i.e., if and only if $\lambda \cdot r < 0$.   ∎ 2.2.17

Corollary 2.2.18 Let $\mathcal{DD} = (D, U, V, \mathcal{I})$ be an affine data dependence and t a valid affine timing function for $\mathcal{DD}$. Then t is finite and bounded if and only if, for all $r \in ray(D)$, $\lambda \cdot r > 0$.   ∎ 2.2.18

Corollary 2.2.19   Let $\mathcal{DD} = (D, U, V, \mathcal{I})$ be an affine data dependence and t a valid affine timing function for $\mathcal{DD}$. Then:

i) for all $r \in ray(D)$ and $z, z' \in D$ such that $z' = z + ar$, with $a \in \mathbf{Z}$, $a \neq 0$, if t is finite then $t(z) \neq t(z')$;

ii) for all $r \in ray(D)$ and $z, z' \in D$ such that $z' = z + ar$, with $a \in \mathbf{Z}$, $a > 0$, if t is bounded below then $t(z) < t(z')$.

$\blacksquare$ 2.2.19

Here is the main result of this section. This theorem is based on the formulation given in [QuVa89]:

Theorem 2.2.20   Let $\mathcal{DD} = (D, U, V, \mathcal{I})$ be an affine data dependence, $\Theta_{\mathcal{I}}$ its dependence mapping and $\Omega_{\mathcal{I}} = range_{\Theta_{\mathcal{I}}}(D)$. Let t be an affine timing function, such that for all $z \in \mathbf{Z}^n$, $t(z) = \lambda \cdot z + \mu$, with $\lambda \in \mathbf{Z}^n$ and with $\mu \in \mathbf{Z}$. Then t is a non-negative, finite, bounded valid timing function for $\mathcal{DD}$ if and only if:

i) for all $v \in vert(\Omega_{\mathcal{I}})$, $\lambda \cdot v > 0$;

ii) for all $r \in ray(\Omega_{\mathcal{I}})$, $\lambda \cdot r \geq 0$;

iii) for all $v \in vert(D)$, $\lambda \cdot v + \mu \geq 0$;

iv) for all $r \in ray(D)$, $\lambda \cdot r > 0$.

PROOF: First we prove that conditions i)-iv) implies that t is a non-negative, finite, bounded, valid timing function for $\mathcal{DD}$. Condition i) and ii) implies that t is valid (see Lemma 2.2.15). Conditions iii) and iv) implies that t is non-negative. In fact, let $vert(D) = \{v_1, \ldots, v_m\}$ and $ray(D) = \{r_1, \ldots, r_p\}$. Then, for all $z \in D$, z can be expressed as:

$$z = \sum_i a_i v_i + \sum_j b_j r_j,$$

with $a_i, b_j \geq 0$ and $\sum_i a_i = 1$. Because of conditions iii) and iv), $\sum_i a_i(\lambda \cdot v_i + \mu) + \sum_j b_j \lambda \cdot r_j \geq 0$. Therefore:

$$\sum_i a_i(\lambda \cdot v_i + \mu) + \sum_j b_j \lambda \cdot r_j =$$

$$= \sum_i a_i \lambda \cdot v_i + \sum_i a_i \mu + \sum_j b_j \lambda \cdot r_j =$$

$$= \lambda \cdot \left(\sum_i a_i v_i + \sum_j b_j r_j \right) + \mu =$$

$$= \lambda \cdot z + \mu = t(z) \geq 0.$$

Finally, condition iv) implies that t is finite and bounded (see Lemma 2.2.17).

Now we prove that if t is a non-negative, finite, bounded, valid timing function for $\mathcal{DD}$ then conditions i)-iv) hold. t valid implies conditions i) and ii) (see Lemma 2.2.15). t non-negative implies condition iii). In fact, as $v \in vert(D)$, in particular, v is an element of D; then $t(v) \geq 0$, i.e., $\lambda \cdot v_i + \mu \geq 0$. Finally, t finite and bounded implies condition iv)(see Lemma 2.2.17).   ■ 2.2.20

Condition ii) of the theorem implies that if $\Omega_{\mathcal{I}}$ contains a line with direction $l \in \mathbf{Z}^n$ ($l \neq 0$), then[10] $\lambda \cdot l = 0$. The following correspondence exists between the conditions of Theorem 2.2.20 and the properties of t:

– Conditions i) and ii) correspond to the validity of t. As polyhedral convex sets are finitely generated, i) and ii) imply that for all $z \in D$, $\lambda \cdot \Theta_{\mathcal{I}}(z) > 0$, which in turn, implies that for all $z \in D$, $t(\mathcal{I}(z)) < t(z)$. Hence, $t(n) < t(n')$, for all pairs of nodes (n, n') of the data dependence graph of $\mathcal{DD}$. As a corollary, the validity of t can be expressed by the condition $\lambda \cdot \Theta(z) > 0$, for all $z \in D$. Also, if there exists a valid affine timing function, then $\Omega_{\mathcal{I}}$ does not contain two non-null vectors d, d' such that $d' = -cd$, with $c > 0$.

– Conditions iii) and iv) correspond to the non-negativity of t over D. This comes from the property of finite generation of polyhedral sets applied to the domain D; and

– Condition iv) corresponds to the finiteness and boundedness of t. In particular, t is finite if and only if $\lambda \cdot r \neq 0$, for all $r \in ray(D)$.

Example 2.2.21   Consider the affine data dependence $\mathcal{DD} = (D, U, V, \mathcal{I})$, with $D = \{(i, j) \mid 1 \leq j \leq n, i \in \mathbf{Z}\}$, where n is a constant in $\mathbf{N}$, and $\mathcal{I}(i, j) = (2i, j - 1)$. The corresponding data dependence graph is depicted in Fig. 2.7 a). Note that D is finitely generated, for instance, by the points

---

[10]If l is the direction of a line of $\Omega_{\mathcal{I}}$, then there exist two rays in $\Omega_{\mathcal{I}}$ with direction l and $-l$, respectively. From condition ii), it follows that $\lambda \cdot l = 0$.

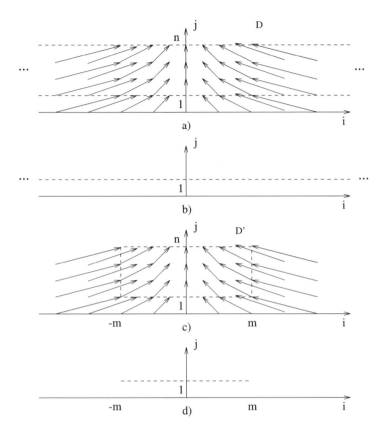

Fig. 2.7. Data dependence $\mathcal{DD}$: a) data dependence graph; b) dependence domain. Data dependence $\mathcal{DD}'$: c) data dependence graph; d) dependence domain.

$(0, 1)$ and $(0, n)$ and directions $(1, 0)$ and $(-1, 0)$. The dependence mapping is $\Theta_{\mathcal{I}}(i, j) = (-i, 1)$ so that $\Omega_{\mathcal{I}}$ is finitely generated by the point $(0, 1)$ and the directions $(1, 0)$ and $(-1, 0)$ (see Fig. 2.7 b)).

A valid affine timing function for $\mathcal{DD}$ is that given by $\lambda = (0, 1)$, which satisfies conditions i) and ii) of Theorem 2.2.20. Note, however, that any affine timing function defined by λ is not finite, as λ does not satisfy the inequality $\lambda \cdot r \neq 0$ for all $r \in ray(D)$. Indeed, for this data dependence there is no valid and finite affine timing function as the constraints for the two properties are in contradiction (as the infinite directions of D and $\Omega_{\mathcal{I}}$ are the same). Now consider the data dependence $\mathcal{DD}' = (D', U, V, \mathcal{I})$, with bounded domain $D' = \{(i, j) \mid 1 \leq i \leq m, 1 \leq j \leq n\}$, where

n and m are constant in $\mathbf{N}$. The corresponding data dependence graph is given in Fig. 2.7 c). Note that D' is finitely generated by its vertices $(-m, 1), (-m, n), (m, 1)$ and (m, n), and $\Omega'_{\mathcal{I}}$, by the vertices $(-m, 1)$ and $(m, 1)$ (see Fig. 2.7 d)). Then $\lambda = (0, 1)$ satisfies all the conditions of Theorem 2.2.20.   ■ 2.2.21

For a uniform data dependence, the dependence domain reduces to a singleton set and conditions i) and ii) of Theorem 2.2.20 reduce to the simpler condition i) of the following corollary (corresponding to a result in [Qui84]):

Corollary 2.2.22   Let $\mathcal{DD} = (D, U, V, \mathcal{I})$ be a uniform data dependence, with index mapping $\mathcal{I}(z) = z + b$, for all $z \in \mathbf{Z}^n$, and $b \in \mathbf{Z}^n$. Let t be an affine timing function, such that for all $z \in \mathbf{Z}^n$, $t(z) = \lambda \cdot z + \mu$, with $\lambda \in \mathbf{Z}^n$ and $\mu \in \mathbf{Z}$. Then t is a non-negative, finite, bounded valid timing function for $\mathcal{DD}$ if and only if:

   i) $\lambda \cdot b < 0$;

   ii) for all $v \in vert(D)$: $\lambda \cdot v + \mu \geq 0$;

   iii) for all. $\in ray(D)$: $\lambda \cdot r > 0$.

                                                                        ■ 2.2.22

Note that the above results refer to a single data dependence. In general, as algorithm specifications are systems of recurrence equations, several data dependencies will have to be considered at the same time. The results extend naturally to a system of equations by considering as domain and dependence domain of the system, the smallest polyhedral convex sets containing the union of, respectively, the domains and dependence domains of its equations.

2.2.6   Affine Scheduling and Dependence Cone

Given a data dependence, its *dependence cone* is defined as the smallest convex polyhedral cone containing its dependence domain:

Definition 2.2.23 *[Dependence Cone]* Let $\mathcal{DD} = (D, U, V, \mathcal{I})$ be an affine data dependence with dependence domain $\Omega_{\mathcal{I}}$. The dependence cone $\Theta^*_{\mathcal{I}}$ of $\mathcal{DD}$ is $\Theta^*_{\mathcal{I}} = cone(\Omega_{\mathcal{I}})$, where $cone(\Omega_{\mathcal{I}})$ is the smallest convex polyhedral cone containing $\Omega_{\mathcal{I}}$.   ■ 2.2.23

In the previous section we have established necessary and sufficient conditions to the existence of a valid affine timing function based on the generators of the dependence domain of an affine data dependence. Here we discuss a sufficient condition for the existence of a valid affine timing function which is based on its dependence cone.

The existence of a valid affine timing function may be related to the pointedness of $\Theta_{\mathcal{I}}^*$. If $\Theta_{\mathcal{I}}^*$ is pointed, then there exists a hyperplane with a non-null, integer normal vector λ such that: $\Theta_{\mathcal{I}}^*$ is entirely contained in one of the half-spaces defined by the hyperplane; and $\lambda \cdot z > 0$ for all z in $\Theta_{\mathcal{I}}^*$. As, by definition, $\Omega_{\mathcal{I}}$ is contained in $\Theta_{\mathcal{I}}^*$, then, in particular, $\lambda \cdot z > 0$ for all z in $\Omega_{\mathcal{I}}$. Therefore, λ defines a (family of) valid affine timing function(s) for the data dependence $\mathcal{DD}$. The formal result is stated in the following proposition:

Proposition 2.2.24 Let $\mathcal{DD} = (D, U, V, \mathcal{I})$ be a data dependence and $\Theta_{\mathcal{I}}^*$ its dependence cone. If $\Theta_{\mathcal{I}}^*$ is pointed then there exists a valid affine timing function for $\mathcal{DD}$.

PROOF: The cone $\Theta_{\mathcal{I}}^*$ is, by definition, a convex polyhedral set. If $\Theta_{\mathcal{I}}^*$ is pointed then it contains no lines, and is finitely generated by its extremal rays. Let $\{r_1, \ldots, r_m\}$ be the set of the extremal rays of $\Theta_{\mathcal{I}}^*$. A valid affine timing function for $\mathcal{DD} = (D, U, V, \mathcal{I})$ is determined by any non-null $\lambda \in \mathbf{Z}^n$ such that $\lambda \cdot r_i > 0$ for all extremal rays r_i. At least one such λ exists because of the separation theorem (see Appendix C).

By definition of $\Theta_{\mathcal{I}}^*$, for all $z \in D$, $\Theta_{\mathcal{I}}(z) \in \Theta_{\mathcal{I}}^*$. Hence, for all $z \in D$, $\Theta_{\mathcal{I}}(z)$ can be written as a positive combination of the extremal rays of $\Theta_{\mathcal{I}}^*$, i.e., $\Theta_{\mathcal{I}}(z) = \sum_i a_i r_i$, with $a_i \geq 0$ and not all $a_i = 0$. Therefore, for all $z \in D$,

$$\lambda \cdot \Theta_{\mathcal{I}}(z) = \lambda \cdot \left(\sum_i a_i r_i\right) =$$

$$= \sum_i a_i \lambda \cdot r_i > 0.$$

Therefore, any timing function defined as $t(z) = \lambda \cdot z + \mu$, for some $\mu \in \mathbf{Z}$, defines a valid timing function for $\mathcal{DD}$. ∎ 2.2.24

Note that pointedness of $\Theta_{\mathcal{I}}^*$ gives only a sufficient condition for the existence of a valid timing function. That it is not necessary can be seen by considering, for instance, the data dependence of Example 2.2.21, which admits a valid (but not finite) affine timing function even though its dependence cone is not pointed (such a cone is $\Theta_\mathcal{I}^* = cone(\{(-1,0),(0,1),(1,0)\})$, corresponding to the dependence domain in Fig. 2.7 b)).

As we will see, the main advantage of this condition is that it can be exploited even in those cases when the dependence domain is not a convex polyhedron. It is this weaker condition that we will use in the remainder of this work.

2.3 Regularisation

One of the requirements of regular array design is that processing elements are locally and uniformly connected. This was motivated originally by technological consideration. If the design is implemented as a VLSI circuit, long wiring is too expensive and introduces undesirable delays in signal propagation. On the other hand, a regular communication topology guarantees simple and cost-effective layouts. Although these requirements can be partially relaxed if regular arrays are implemented in software, regularity remains one the basic feature of regular arrays as a model of computation.

Because of the relationships between data dependence structure of the specification and communication structure of the design, notions of uniformity and locality can be formulated on data dependencies. We have already formally defined what we intend by a uniform data dependence, and we can think of a local data dependence as a data dependence under which only nearest neighbour computation points are related. The lack of regularity of a data dependence has to be addressed before the mapping onto a regular array design is feasible. We use the term *regularisation* to refer to the set of techniques which can be used to improve the uniformity and locality of data dependencies.

As discussed in Chapter 1, regular array synthesis is based on the specification of algorithms as systems of equations whose data dependencies conform to a particular syntax, and, historically, recurrences with uniform data dependencies were the first class of recurrences to be considered, for their natural correspondence with regular arrays. In fact, any regular array can be expressed as a system of uniform recurrences (as proved by Rao in [Rao85]) and uniform recurrences always map to regular array designs under linear space-time mappings. Also, we have already mentioned that in order to facilitate the specification of algorithms, more general classes of recurrences have been introduced together with regularising transformations for their manipulation into systems of uniform recurrences. These new recurrences are those characterised by affine data dependencies, and constitute the most general type of data dependencies which is considered in regular array synthesis at present.

The main goal of the work presented in this book is the extension of synthesis techniques to more general classes of problems. This will be formally addressed in the following chapters. In this section, however, we intend to discuss in general terms the basic issues and properties which are involved in the regularisation of non-uniform data dependencies. We will capitalise on these properties in the development of the regularisation techniques of the following chapters.

2.3.1 Decomposition and Uniformisation

Regularisation involves replacing a non-uniform data dependence with a set of data dependencies which exhibit improved uniformity and locality. Although regularisation mainly refers to transformations of the data dependence graph, it can be described in terms of *pipelining* and *routing* of the data flow between computation points. This terminology can be justified if we think of a data dependence graph as a signal flow graph (this is always possible - consider the trivial allocation function in Section 2.1.3). Also, regularisation has the ultimate goal of producing specifications for which a regular array design is guaranteed under a space-time mapping. Hence, ultimately the application of regularisation techniques is reflected in the improved uniformity and locality of the connection topology of the design. Early forms of regularisation were actually defined directly on the signal flow graph rather than the data dependence graph (see, e.g., the work on retiming in [KuLe80, LeSa83, SYKun88]). The appeal of this terminology is that of providing an intuitive picture of the effect of the transformations, which certainly helps in the understanding of the mathematics.

In general, we associate the term pipelining with a particular direction vector in the lattice space, and we expect the data to be propagated in a regular fashion through the computation points in that direction. On the other hand, we associate the term routing with a set of direction vectors in the lattice space, used to define a sequence of directions used for the propagation of the data among computations. Such directions contribute to the definition of routing paths on the nodes of the dependence graph. A regular routing scheme is one such that the propagation of the data flow in

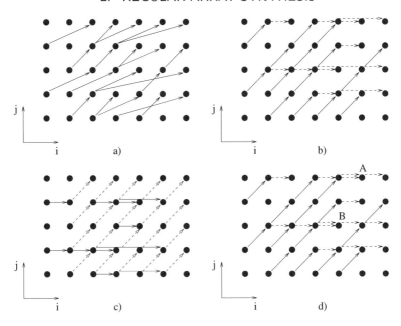

Fig. 2.8. Regularisation: a) a generic data dependence; b) and c) possible decompositions; d) a possible uniformisation.

each of the routing directions is pipelined.

Given a non-uniform data dependence, its regularisation may be achieved by selecting a number of routing directions in the lattice space and pipelining the data flow in each of those directions. We illustrate these two activities on the data dependence graph depicted in Fig. 2.8 a).

The basic requirements for the choice of routing directions, is that each data dependence vector can be expressed as an integer combination of those directions. For instance, any unimodular basis of the space will meet this requirement. The integer combination can be used for the definition of a routing path which replaces the dependence vector, simply by fixing an ordering of the routing vectors and considering the summands of the combination in the sequence established by the ordering. In our example, let us choose as routing directions the vectors $(1,1)$ and $(1,0)$. These vectors constitute a unimodular basis in $\mathbf{Z}^2$, therefore any data dependence vector in Fig. 2.8 a) can be rewritten as a combination of $(1,1)$ and $(1,0)$ with suitable integer coefficients. By considering $(1,1)$ and $(1,0)$ in this order, the data dependence vectors of Fig. 2.8 a) may be replaced by the routing paths of Fig. 2.8 b), where solid lines correspond to the direction $(1,1)$

and dashed lines to $(1, 0)$. Reversing the ordering of the routing directions produces the paths of Fig. 2.8 c). Note that one of the effects of this substitution is that more points of the lattice space are involved in the transfer of the data among computations. Hence, from a point-to-point connection represented by a data dependence vector, we define a path of lattice points through which the data is transferred. The end points of the path are the pair of points initially related by the dependence vector.

In the following chapters, we will use the term *decomposition* for the techniques which allow the definition of these routing paths for various classes of non-uniform data dependencies. In general, decomposition will also allow us to consider separately the sub-paths generated in each of the routing directions (hence the name we have chosen for the transformation).

Once the routing paths have been established by a selection of the routing directions and their ordering, the data flow can be pipelined along the paths. In our example, the routing paths of Fig. 2.8 b) are transformed into those of Fig. 2.8 d), by pipelining the data according to the direction of $(1, 0)$. Note that the result of this transformation is a set of more dense routing paths, which involve nearest neighbour points in each of the routing directions. The name which we have chosen for this transformation is *uniformisation*, as, in general, it corresponds to the definition of a set of uniform data dependencies.

2.3.2 Data Conflicts and Data Broadcasts

A *data conflict* is an attempt by distinct pieces of data to share the same communication channel during the same time cycle. A data conflict is avoided by providing distinct communication channels for each of these pieces of data. If the array design is realised in hardware, because of the cost involved, the amount of data conflicts needs to be contained. Also, if the level of data conflicts increases with the size of the problem, the array design is not scalable. This is a strong limitation for hardware implementation as it would involve the replacement of the whole piece of hardware every time a problem of larger size needs to be solved. Therefore, array designs characterised by data conflicts which are dependent on the size of the problem should be avoided.

Data conflicts may be generated by *overloading* the dependence graph of a specification. The nodes of a dependence graph become overloaded when the pipelining of the data flow according to a certain direction produces several routing paths which cross the same computation points. Then distinct communication channels at the corresponding processing elements need to be provided in order to avoid conflicts. If the level of overloading is dependent on the problem size, the level of conflicts exhibits the same dependence. Therefore, a major constraint in the selection of routing directions is to avoid directions of the lattice space which yield problem size dependent overloading of the data dependence graph. Let us consider once again the data dependence graph of Fig. 2.8 a). The choice we made of routing directions and their ordering resulted in the graphs in Fig. 2.8 b) and d). Note that some of the nodes of the graph in Fig. 2.8 d) are overloaded (those labelled as A and B).

The appearance of overloading can be related to the presence of *data broadcasts*. In terms of array design, a data broadcast indicates that the same data is shared by several processing elements during the same time cycle, and therefore has to be transferred from some source processing element to all the recipients with a direct communication. Data broadcasts usually involve non-local communication between processing elements and in regular array design are always replaced by a pipelining of the data among neighbour recipients. In a specification, a data broadcast corresponds to a one-to-many data dependence relation, that is several nodes of the dependence graph depend on the same node. In Fig. 2.8 a) several data broadcasts are present. When compared with Fig. 2.8 d) the overloading appears exactly on those paths replacing data broadcasts. A more precise way of defining a data broadcast is as the data dependence relation defined by a non-injective index mapping: if several points depend on the same point, that point is their image under the index mapping, and the mapping is not injective. Therefore, overloading the nodes of the graph can be avoided by avoiding the definition of data broadcasts.

It is this important principle which is exploited in the definition of the regularisation techniques of the next chapters: we select routing directions such that their routing paths correspond to data dependencies defined by injective index mappings. Indeed the question to address is whether this selection is always possible. Fortunately, in general the answer is affirmative, but it may require an increase in the number of dimensions of the specification. For instance, let us consider, once again, the graph of Fig. 2.8 b). Overloading can be avoided by adding a third dimension to the problem and producing the graph of Fig. 2.9. In the figure, white nodes denote the extra routing points in the 3-dimensional space. Note that not only an increase of dimensionality is required, but also new routing directions are used (the order of the paths, however, remains unchanged).

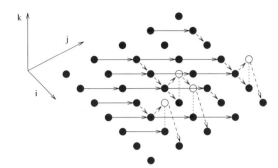

Fig. 2.9. Non-overloaded 3-dimensional data dependence graph.

2.3.3 Regularisation and Dependence Cone

Because of the relation between affine timing functions and pointed dependence cones, a condition for the preservation of affine scheduling by regularisation may be formulated in terms of the preservation of pointedness of the dependence cones. In particular, the routing directions need to be chosen as vectors of the lattice space which, together with the original dependence cone of the specification, generate a pointed polyhedral convex cone. This is possible, in general, but may require an increase of the dimensionality of the space. The formulation of this condition for various types of data dependencies will be provided in the following chapters.

2.3.4 Substitution of a Data Dependence

The definition of a routing scheme which replaces a data dependence finds a mathematical expression in the function composition of index mappings. It is not limited, however, to such a composition as the computation domains and variables of the data dependencies have also to be taken into account. The definition of a routing scheme may be described as follows. Let $\mathcal{DD} = (D, U, V, \mathcal{I})$ be a data dependence describing the dependence relation between the instances of variables U and V on the domain D. In particular, for all $z \in D$, $U(z)$ is data dependent on $V(\mathcal{I}(z))$. In other words, the value of V at the computation point $\mathcal{I}(z)$ has to be transferred to z in order to enable the computation of U. $\mathcal{DD}$ implies that there exists a direct data communication from $\mathcal{I}(z)$ to z. A routing scheme for $\mathcal{DD}$ defines a communication path from $\mathcal{I}(z)$ to z via a number of intermediate points of the lattice space. The routing scheme has to guarantee the following two requirements: that for each $z \in D$ the communication path from $\mathcal{I}(z)$ to z is finite; and that the value $V(\mathcal{I}(z))$ is transferred among neighbour points on the communication path from $\mathcal{I}(z)$ to z via a set of routing variables.

In order to define the routing scheme, a system of recurrence equations is introduced which defines the routing variables, and the original data dependence is replaced by a new data dependence which is defined on such variables. Let us illustrate the transformation in the following example.

Example 2.3.1   Consider the data dependence $\mathcal{DD} = (D, U, V, \mathcal{I})$, such that $D = \{(i,j) \mid i \geq 1, 1 \leq j \leq m\}$, for some m in $\mathbf{N}$, and $\mathcal{I}(i,j) = (0,j)$. $\mathcal{I}$ defines the data broadcasts sketched in Fig. 2.10 a), that is for each point in D, $U(i,j)$ depends on the value $V(\mathcal{I}(i,j))$. For each point of D, we want to replace any such data broadcast with a pipelined propagation as illustrated in Fig. 2.10 b). This is achieved by replacing $\mathcal{DD}$ by the data dependence

$$\mathcal{DD}' = (D, U, R, \mathcal{I}_0)$$

where $\mathcal{I}_0(i,j) = (i,j)$ and R is the routing variable defined by the equations:

$$
\begin{aligned}
\mathbf{E}_1 &= (D_1, R, R, id, \mathcal{I}_1) \\
\mathbf{E}_2 &= (D_2, R, V, id, \mathcal{I}_1)
\end{aligned}
$$

with $\mathcal{I}_1(i,j) = (i-1,j)$, $id(a) = a$, and

$$
\begin{aligned}
D_1 &= \{(i,j) \mid i \geq 2, 1 \leq j \leq m\} \\
D_2 &= \{(i,j) \mid i = 1, 1 \leq j \leq m\}.
\end{aligned}
$$

Now we have that for each point $(i,j) \in D$, $U(i,j)$ depends on $R(\mathcal{I}(i,j)) = R(i,j)$. In turns $R(i,j)$ is defined as:

- for $i = 1$ (that is in D_2), equal to $V(0,j)$;

- for $i > 1$ (that is in D_1), equal to $R(i-1,j) = R(i-2,j) = \ldots = R(1,j) = V(0,j)$.

By induction, we can show that, for all $(i,j) \in D$, $\mathcal{I}(i,j) = \mathcal{I}_1^i \circ \mathcal{I}_0(i,j)$, where $\mathcal{I}_1^i$ denotes the composition $\mathcal{I}_1 \circ \mathcal{I}_1 \circ \ldots \circ \mathcal{I}_1$, i times. Therefore we have that for each point in D, $U(i,j)$ depends:

$$
\begin{aligned}
R(\mathcal{I}_0(i,j)) &= R(\mathcal{I}_1^{i-1} \circ \mathcal{I}_0(i,j)) = \\
&= V(\mathcal{I}_1^i \circ \mathcal{I}_0(i,j)) = V(\mathcal{I}(i,j)).
\end{aligned}
$$

This shows that the new system defining a pipelining propagation of the data is equivalent to the system defining the initial broadcast.   ∎ 2.3.1

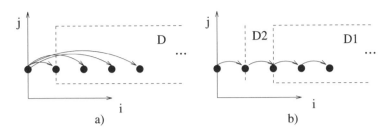

Fig. 2.10. Substitution of data broadcasts with pipelined propagation.

2.4 A Brief Survey

The basic ideas behind regular array design can be related (see [Meg92]) to the theory of cellular automata, whose foundations were established by John von Neumann [VonN66] in the early 60s. Cellular automata were initially developed for the study of the evolution of biological systems, although several applications in mathematical and physical sciences have successively been developed. Cellular automata deal with large (possibly infinite) collections of interconnected finite state automata and, hence provide a framework for the investigation of systems characterised by homogeneous and scalable components. (An overview of the theory and application of cellular automata can be found in [Wol86].)

The formal development of regular array synthesis began in the late 60s with the work by Karp, Miller and Winograd [Ka-et-al67], who proposed the idea of expressing (classes of) iterative algorithms as systems of uniform recurrence equations (UREs). In their work, the computability of the algorithm was related to properties of the reduced dependence graph of the equations, and necessary and sufficient conditions for the existence of a scheduling were stated.

While the work by Karp *et al.* is based on a functional representation of an algorithm, the work by Lamport [Lam74], may be considered as its imperative counterpart. His work is based on the observation that for large classes of numerical algorithms, most of the computation is devoted to the execution of loops, in particular FORTRAN-like for-loops. Under a number of assumptions on the form of the loops (such as the absence of input/output operations or of control transferred outside the loop), Lamport proposed a systematic rewriting of sequential nested for-loops into concurrent loops, based on the analysis of the data dependencies. He also introduced the notion of equitemporal hyperplanes for the scheduling of the computations, a notion which has been extensively exploited in regular array synthesis. His work has been also most influential in the subsequent development of parallel compilers.

In the early 70s, data flow computing [Ada68, Ada70, DeWe77, Den80] was defined for the maximal exploitation of parallelism as an alternative to conventional control flow computers. The essential idea of data flow computing is that of enabling the execution of an instruction as soon as its operands become available. In other words, computations are driven by data availability rather than explicit control.

In the early 80s, systolic arrays were introduced by H.T.Kung and C.E. Leiserson [KuLe80]. The term array indicates their structural affinity with array processors, while the term systolic describes their behaviour using the human circulatory system as a metaphor (signals are rhythmically "pumped" among processing elements). Systolic arrays are structurally simple, regular and modular and, typically, are realised through replication and local interconnection of simple processing elements which perform basic operations. Multiprocessing and pipelining are principles of systolic behaviour ensuring high performance with low memory and input/output bandwidth: the same data can be propagated among neighbouring processing elements to be, thus, reused by the recipients. Also, parallelism is synchronous and decentralised: computations occur in lockstep, with signals representing both data and control information. Although systolic arrays were initially designed for hardware implementation as VLSI circuits, the principles of systolic computation are now considered more generally as a paradigm for parallel processing, and in synthesis methods they are retained as a computational model.

Asynchronous regular arrays, so-called wavefront arrays , were introduced by S.Y.Kung et al. in the early 80s [Ku-et-al81, Ku-et-al82]. Wavefront arrays are based on the principles of data flow computing. The name wavefront is evocative of the way their computations proceed, which resembles a wave propagation. A major advantage of wavefront arrays over systolic arrays is that they do not require global synchronisation.

An automata and complexity theory for systolic computing, known as systolic automata theory, developed through the 80s and early 90s [Cu-et-al83, Cu-et-al84, Gru84, FaNa88, Gru90], mainly dealing with a restricted class of problems which have systolic solutions as linear array, trellis or tree-like forms.

From an applicative point of view, a vast literature exists on systolic arrays for the solution of numerical and non-numerical problems. Systolic arrays have become a popular form of parallel computing, and collections of systolic algorithms can be found, for example, in [QuRo91, Eva91, Meg92].

Initially, array designs were derived manually and in an *ad hoc* fashion. Attempts at coding the expertise acquired in the development of systolic algorithms into a methodology with mechanised support started in the middle of the 80s. Between 1983 and 1985 a number of independent contributions defined the foundations of regular array synthesis. Moldovan [Mol83] introduced the notion of space-time mapping of the data dependence graph as a pair of linear functions. He also defined the concept of dependence mapping (rather than simply a dependence vector as it is for UREs), hence defining a new type of recurrence equations for the specification of algorithms. We refer to such recurrences as affine recurrence equations (AREs), a name introduced later by Rajopadhye and Fujimoto [RaFu90]. Quinton [Qui84] proposed the detection of a scheduling for system of UREs as the solution of an (integer) linear optimisation problem. Cappello and Steiglitz [CaSt84] focused on the geometric representation of an algorithm and its data dependencies as a graph in a Euclidean lattice space together with sets of linear transformations of such a representation. They showed that this geometric framework is a powerful unification tool, as several array designs for the same algorithm correspond to the mapping of data dependence graphs which are linear transforms of the same specification. Miranker and Winkler's contribution [MiWi84] can be seen as emphasising that the mapping problem reduces to a mapping between graphs: the space-time representation of the algorithm and its data flow representation.

Between 1985 and the early 90s, several contributions followed these seminal works on methodology. It would be impractical to enter the details of each of them. To summarise, althogh there were no major revisions of the methodology, particular aspects of regular array synthesis were addressed, resulting in the refinement of the design process to the structure illustrated in Fig. 2.11. This diagram is a more detailed version of that of Fig. 2.1 at the beginning of this chapter, with more steps included in the core of the design process. In the following we provide a brief review of some of the major contributions.

The analysis step, together with the analysis of the data dependencies, also addresses the *computability* of the specification, where a computable specification is a specification which admits a valid timing function. The computability of uniform and affine recurrence equations has been investigated widely [Ka-et-al67, Rao85, DeIp86, DeIp87, Ra-Ka88, SaQu90]. It has

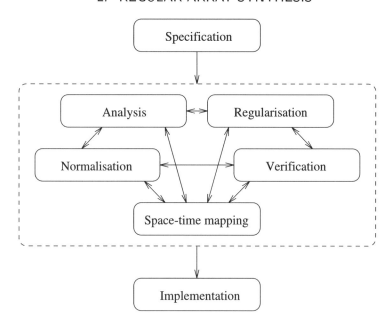

Fig. 2.11. Design process.

been established that if the computation domains are unbounded, the computability of a system of recurrence equations can be reduced to the *halting problem* [BoJe89], and so is undecidable. On the other hand, for uniform recurrence equations with a finite computation set, computability is decidable and corresponds to checking whether the (complete) data dependence graph is acyclic.

A correspondence between systolic arrays and algorithms specified by recurrence equations was formally established by Rao in his doctoral thesis in 1985 [Rao85]. In particular he showed that if an algorithm can be implemented as a systolic array then it can be expressed as a Regular Iterative Algorithm (RIA), an extension of UREs with finite conditional branches in each recurrence equation. Li and Wah [LiWa85] treated the derivation of a systolic design from a specification as an *optimisation* problem, based on a set of design parameters such as the velocity of data flows, the spatial distribution of data, or the periods of computation. Similarly, M. Chen [Che86] treated the design process as an optimisation problem applied to algorithms specified in the parallel programming language Crystal [Che86b]. Her work included an initial investigation into aspects of the design such as regularisation, synthesis of control signals and mapping to fixed-size architectures.

Delosme and Ipsen [DeIp86] concentrated on computability issues and extended the work of Karp *et al.* [Ka-et-al67] to systems of AREs. Based on the work by Delosme and Ipsen, Yaacoby and Cappello [YaCa88] provided necessary and sufficient conditions for the existence of an affine scheduling and a procedure to construct a scheduling vector. Similarly, Rajopadhye and Fujimoto [RaFu90] extended the work of Quinton [Qui84] to AREs, defining a linear optimisation problem for the automatic derivation of an affine scheduling.

Early work on *regularisation* includes the retiming of a signal flow graph, for example, by Leiserson and Saxe [LeSa83]. Subsequent regularisation techniques were defined by several authors, mainly for the removal of data broadcast in AREs. As data broadcasts correspond to non-injective index mappings, for affine data dependencies, they can be related to rank deficient matrices (defining the linear part of the index mapping), and simple forms of regularisation can be provided based on the selection of pipelining directions in the null space of such matrices. This approach was taken by Fortes and Moldovan [FoMo84] and Rajopadhye and Fujimoto [RaFu87, Raj89]. Wong and Delosme [WoDe92] proposed more general forms of regularisation of data broadcasts based on the selection of routing vectors as the elements of canonical and non-canonical basis of the lattice space. Regularisation techniques for more general forms of non-uniform AREs (i.e., not limited to data broadcasts) were proposed by Quinton and Van Dongen [QuVa89], via a combination of pipelining and routing. They also defined a new class of recurrences, generalising AREs with linear size parameters. The work presented in this book on the regularisation of non-affine types of data dependencies has developed from their approach.

The systematic *derivation of control signals* from conditional expressions was first addressed by M. Chen [Che86] in 1986. In this work she proposed methods to replace conditionals of a recurrence equation by control signals pipelined from the boundary of the array. Radjopadhye and Fujimoto [RaFu87] also proposed a systematic pipelining of conditional expressions. However, they considered a more restricted class of conditional expressions which arise because of the application of regularisation techniques to AREs. The characterisation of such conditionals resulted in the definition of the class of so-called conditional uniform recurrence equations (CUREs). Teich and Thiele in [TeTh91] adopted an approach similar to Chen's for what they defined as piecewise regular iterative algorithms (an extension of Rao's RIAs with affine data dependencies). Their formalism is based on Chandry and

Misra's UNITY [ChMi88]. Finally, Xue [XuLe92, Xue92] proposed a more general method that also extends to space-time mappings of a dependence graph onto fixed-size array designs.

Regular array synthesis aims at producing optimal array designs based on the assumption that unbounded computational resources are available. For more realistic situations in which this assumption cannot be made, *partitioning* techniques have been developed which allows the mapping of, possibly, multi-dimensional algorithms onto fixed-size lower-dimensional array designs. The development of partitioning techniques is not confined to regular array design, but is part of the more general problem of parallel code generation from sequential code (see, e.g., [IrTr88, AnIr91, Fea92b]). The main goal of partitioning methods is to operate a compression of the design. This can be achieved either at the data dependence level, by defining clusters of computations to allocate onto the same processor, or at the signal flow graph level, by merging clusters of cells into a single (super-) processor. Partitioning techniques are usually related to the use of multi-dimensional schedules (see, e.g., [Fea92b]), i.e., the timing function is a mapping between multi-dimensional lattice spaces. Work on partitioning for regular arrays started in 1986 with a contribution by Moldovan and Fortes [MoFo86]. Partitioning techniques were classified by Darte [Dar91] into Locally Parallel Globally Sequential (LPGS) and Locally Sequential Globally Parallel (LSGP). In LPGS partitioning, each partition is a block of parallel computations, while the blocks are processed sequentially. This class includes the work by Moldovan and Fortes [MoFo86] and Bu *et al.* [Bu-et-al90]. In LSGP partitioning, each partition is a sequential block of computations and blocks are executed in parallel. Darte's work [Dar91] belongs to this class. Independent partitioning was proposed by Shang and Fortes [ShFo92b], where independence means that no communication is needed between the blocks of the partition, while X. Chen and Megson [MeCh94] related partitioning to code generation for parallel platforms (in particular, transputers), by exploiting the idea of canonical dependencies derived from a positive expressive basis.

Verification refers to the formal proof of some correctness properties of a regular array design. Early work on verification, undertaken since 1983, includes: the approach by M. Chen [Che83], based on systems of recurrence equations and fixed point induction; the algebraic approach by Kung and Lin [KuLi84]; and the approach based on the solution of systems of differential equations by Melhem and Rheinboldt [MeRh84]. In 1986, Hennessy [Hen86] used process algebras for the specification of a systolic circuit and fixed point induction to derive an implementation from the specification. With work started in 1988, Thompson and Tucker [ThTu88, ThTu91, ThTu94] have developed formal specification and verification techniques for *Synchronous Concurrent Algorithms (SCA)*, of which systolic arrays are a particular case. Their method is based on many-sorted universal algebras, primitive recursion and equational logic. Work towards automatic verification systems for systolic arrays includes: the approach by Abdulla [Abd90, Abd92], which addresses a general model for the description and verification of systolic circuits over arbitrary algebras (he provides completely automatic verification for subclasses of systolic circuits [Abd92]); and the work by Ling and Bayoumi [LiBa94], which defines a systolic temporal arithmetic (based on temporal logic) for the specification and verification of systolic designs at the array level (for which they provide a Prolog-based verifier).

As the automation of the design process is one of the basic objectives of regular array synthesis, a number of support software tools and environments have been developed. Among the major contributions in the form of Computer-Aided Design (CAD) tools, we may recall DIASTOL [Ga-et-al87] and its more complete version Alpha du Centaur (AdC) [Ga-et-al88], ADVIS (Automatic Design of VLSI Systems) [Mol87], PRESAGE, [VanD88], DE-COMP [VeCr91] and SADE (Systolic Array Design Environment) [MeCo91]. In general, these tools support some of the basic steps of regular array synthesis, such as the generation of the data dependence graph and its (optimal) space-time mapping. Languages for the initial specification of an algorithm vary from case to case, and include specialised languages based on recurrence equations or restricted forms of nested for-loops. Early tools accept as inputs only uniform problems, while more recent developments can be used also for the synthesis of affine problems. Often graphical interfaces have been developed for the representation of the data dependence graph as well as the animation of the array design through snapshots of computations. More sophisticated tools also include parallel code generation. Compilers for systolic and regular programs have also been developed. The first systolic compiler was developed by H.T.Kung *et al.* in the early 80s, for the CMU Warp machine [KuWe85, An-et-al87], a systolic array computer of linearly interconnected programmable cells. A more general approach to systolic compilation is due to Lengauer *et al.* [HuLe87], in which, from a formally specified program, traces (of operations) are extracted, transformed into parallel traces, and a corresponding systolic design is derived. The design can then be either implemented in hardware or corresponding parallel code can be generated for a target machine [Le-et-al91].

From the recognition that complex algorithms may be defined by combining regular components, a novel approach to regular algorithm design comes from *piecewise regular arrays*. Piecewise regular arrays combine the strength of traditional data dependence analysis and transformations with the exploitation of associativity and commutativity properties of computations. A comprehensive treatise of piecewise regular arrays can be found in [Pla99].

2.5   Summary

In this chapter we have introduced some of the basic concepts of regular array synthesis. In particular, we have outlined the design process and discussed the basic notions of recurrence and input equation, equation system, data dependence and related graphical representations. We have explained how regular arrays can be derived from algorithm specifications through space-time mappings, that is a scheduling of the computations and their placement onto processing elements. We have stated formal properties of space-time mappings, such as the validity, finiteness and boundedness of the scheduling, and the compatibility of scheduling and placement.

We have discussed the advantages of developing synthesis methods in the framework of Euclidean geometry, both theoretically and in the practical development of systematic transformations. In this framework, powerful techniques can be developed by restricting ourselves to affine data dependencies and affine timing and allocation functions, and exploiting the basic properties of linear and affine spaces. Among the major advantages is the possibility for the systematic derivation of optimal affine timing functions. Central to Euclidean synthesis methods is the concept of dependence cone, as a polyhedral convex cone embedding all possible data dependence vectors of an algorithm specification. Its pointedness can be related to the existence of linear schedulings and their preservation by transformations of the specification.

We have discussed at length the rôle of regularisation in the design process, as the development of regularisation techniques is the main theme of this work. We have outlined routing and pipelining, on a small example, as basic transformations for regularisation. We have explained the relation between data conflicts and data broadcasts, and how they can be detected and avoided in regularisation. We have discussed the relation between regularisation directions and dependence cones together with guidelines for the preservation of affine timing functions.

From the discussion, a regularisation scheme has emerged, which will be used in the development of the following chapters. It consists of a syntactic characterisation of classes of non-uniform data dependencies, and their systematic substitution with new data dependencies which exhibit improved locality and uniformity, the target being specifications as systems of uniform recurrence equations. The new data dependencies are defined by selecting regularisation directions in the embedding Euclidean space and defining regular routing systems for transferring data among computation points. We need to provide conditions for the preservation of affine schedulings by relating regularisation directions and (the generators of) data dependence cones. We also need to guarantee the absence of data conflicts.

In the last part of the chapter we have presented a brief survey of the major contributions in regular array synthesis which can be found in the literature.

Chapter 3

Integral Recurrence Equations

In this chapter we introduce *integral recurrence equations* and their systematic regularisation. We will base the definition of integral recurrence on the notion of *integral index mapping*, as an index mapping in $\mathbf{Z}^n$ which is not required to define an affine transformation. The syntactic form of an integral index mapping is an integer combination of a finite set of directions of the lattice space, in which the coefficients are functions from $\mathbf{Z}^n$ to $\mathbf{Z}$. With this form we will be able to establish an explicit relation between the index mapping and a finite set of vectors of the space, which can be exploited for regularisation purposes. Based on this syntactic form we will show that affine data dependencies are particular types of integral data dependencies.

The regularisation techniques that we will define will allow us to transform integral specifications systematically into systems of uniform recurrence equations. This fact implies that ordinary mapping techniques can subsequently be applied for the derivation of regular array designs. We will choose regularisation directions among the direction vectors defining the index mappings. Based on the same vectors, we will give conditions for the existence of affine timing functions and their preservation through regularisation.

The main difficulty in the definition of regularisation techniques for integral data dependencies stems from the necessity of reconciling the existence of polyhedral convex sets and the application of non-affine transformations. Because of the rôle of convexity in Euclidean synthesis, we need to guarantee that all computation domains are convex polyhedra. However, non-affine index mappings, in general, do not preserve convexity. The solution we have chosen is that of enforcing convexity, at the expense of an increased complexity of the regularisation techniques. In particular, we will make use of control variables, where a control variable is an ordinary variable of the specification, whose values are interpreted as control signals rather than data. As in a regular array, control signals (other than clock synchronisation) are entirely distributed, the distinction between data and control variables is rather artificial. However, it emphasises the particular rôle that these variables play in the computation. In our techniques, we will consider enlarged polyhedral convex domains and define control variables in order to identify non-convex subsets of computation points in those domains. Hence, the solution we will adopt introduces some control overhead, but it allows us to recover convexity and remain within the domain of classic regular array synthesis. As we will see in the next chapter, the use of control variables also accounts for a degree of reconfigurability of the array design, which we will exploit for particular classes of dynamic data dependencies.

This chapter is organised as follows. In Section 3.1 we define integral index mappings and data dependencies, and discuss their relation with the affine and uniform cases. Section 3.2 is devoted to regularisation techniques. In particular, we establish a condition for the injectivity of an integral index mapping, and use it for the development of regularisation techniques which guarantee that no size dependent overloading of the dependence graph is generated. The regularisation techniques developed have the form of decomposition and uniformisation techniques. We also define a parametric version of uniformisation, where the parameter is seen as an upper bound on the level of overloading allowed, hence the level of conflicts and corresponding physical resources of the array design can be controlled to some extent. Parametric uniformisation allows for compact designs at the expense of increased communication resources. In Section 3.3, we formally address the fact that the regularisation techniques preserve affine scheduling. In the chapter we will use toy examples to clarify the basic results. We will present more interesting applications of the techniques in Chapter 5.

3.1 Integral Data Dependencies

We define an *integral index mapping* as an index mapping in $\mathbf{Z}^n$, which can be expressed as an integer combination of a finite set of direction vectors in the lattice space. In addition, when only one direction vector is used, we call the index mapping *atomic integral* (or simply *atomic*).

Definition 3.1.1 *[Integral Index Mapping]* Let $\mathcal{I}$ be an index mapping. $\mathcal{I}$ is integral if for all $z \in \mathbf{Z}^n$, $\mathcal{I}(z) = z + \sum_{j=1}^{m} g_j(z)d_j$, where, for $j = 1, \ldots, m$, $g_j : \mathbf{Z}^n \to \mathbf{Z}$ and d_j is a non-null vector in $\mathbf{Z}^n$. In addition, $\mathcal{I}$ is atomic integral if $m = 1$.   ∎ 3.1.1

We call the vectors d_j the generators of $\mathcal{I}$, and the integer functions g_j its coefficients. The form of an integral index mapping of Definition 3.1.1 is not unique, as illustrated in the following example.

Example 3.1.2   Consider the index mapping $\mathcal{I}(i, j, k) = (i - 2^j, j - 1, k - 1)$. $\mathcal{I}$ may be expressed, for instance, by any of the two integral forms:

$$
\begin{aligned}
\mathcal{I}(i, j, k) &= (i, j, k) + 2^j(-1, 0, 0) + 1(0, -1, 0) + 1(0, 0, -1) \\
\mathcal{I}(i, j, k) &= (i, j, k) + 2^j(-1, 0, 0) + 1(0, -1, -1)
\end{aligned}
$$

∎ 3.1.2

In theory, any number of generators and corresponding coefficients can be chosen to express an integral index mapping. In practice, however, only expressions with a number of generators less or equal to the number of dimensions of the lattice space are actually considered. Indeed, this is always possible (trivially, by considering the elements of the standard basis of the space). As we will see, the choice of generators has an impact on the regularisation of the corresponding data dependencies, hence on the complexity of the resulting array design. We will consider this issue in greater detail in Section 3.2.1.

Integral dependence mappings may be defined, similarly to the affine case, as follows:

Definition 3.1.3 *[Integral Dependence Mapping]* Let $\mathcal{I}$ be an integral index mapping. Its dependence mapping $\Theta_{\mathcal{I}}$ is the mapping defined as $\Theta_{\mathcal{I}}(z) = z - \mathcal{I}(z)$, for all $z \in \mathbf{Z}^n$.   ∎ 3.1.3

If $\mathcal{I}$ is an integral index mapping defined as $\mathcal{I}(z) = z + \sum_{j=1}^{m} g_j(z)d_j$, its dependence mapping is $\Theta_{\mathcal{I}}(z) = -\sum_{j=1}^{m} g_j(z)d_j = \sum_{j=1}^{m} g_j(z)(-d_j)$. We call the vectors $-d_j$ the generators of $\Theta_{\mathcal{I}}$ and the integer functions g_j its coefficients. Note that, in general, $\Theta_{\mathcal{I}}$ is not affine. Integral index mappings characterise integral data dependencies:

Definition 3.1.4 *[Integral Data Dependence]* Let $\mathcal{DD} = (D, U, V, \mathcal{I})$ be a data dependence. $\mathcal{DD}$ is integral if $\mathcal{I}$ is an integral index mapping. In addition, $\mathcal{DD}$ is atomic integral if $\mathcal{I}$ is atomic integral.   ■ 3.1.4

The relation between atomic integral and integral data dependencies is similar to that between uniform and affine data dependencies. In particular, atomic integral data dependencies are simple forms of integral data dependencies, and for regularisation purposes we will aim at substituting a generic integral data dependence with a set of corresponding atomic integral data dependencies.

The dependence domain and cone of an integral data dependence can also be defined similarly to the affine case.

Definition 3.1.5 *[Integral Dependence Domain]* Let $\mathcal{DD} = (D, U, V, \mathcal{I})$ be an integral data dependence and $\Theta_{\mathcal{I}}$ the dependence mapping defined by $\mathcal{I}$. The dependence domain $\Omega_{\mathcal{I}}$ of $\mathcal{DD}$ is $\Omega_{\mathcal{I}} = \Theta_{\mathcal{I}}(D)$.   ■ 3.1.5

As, in general, $\Theta_{\mathcal{I}}$ is not affine, the dependence domain $\Omega_{\mathcal{I}}$ is not a convex polyhedral set.

Definition 3.1.6 *[Integral Dependence Cone]* Let $\mathcal{DD} = (D, U, V, \mathcal{I})$ be an integral data dependence with dependence domain $\Omega_{\mathcal{I}}$. The dependence cone $\Theta_{\mathcal{I}}^{*}$ of $\mathcal{DD}$ is $\Theta_{\mathcal{I}}^{*} = cone(\Omega_{\mathcal{I}})$.   ■ 3.1.6

Because of the relation between pointed dependence cones and affine scheduling discussed in Section 2.2.6, in the following we always assume that $\Theta_{\mathcal{I}}^{*}$ is pointed.

Example 3.1.7   Consider the data dependence $\mathcal{DD} = (D, U, V, \mathcal{I})$ with domain $D = \{(i, j, k) \mid 1 \leq i, j \leq p, k = 1\}$, for some $p \in \mathbf{N}$, and index mapping $\mathcal{I}(i, j, k) = (i, j, k) + 2^{j}(-1, 0, 0) + (0, -1, -1)$. Its dependence mapping is $\Theta_{\mathcal{I}}(i, j, k) = 2^{j}(1, 0, 0) + (0, 1, 1)$. Its dependence domain is $\Omega_{\mathcal{I}} = \{(2^{j}, 1, 1) \mid 1 \leq j \leq p\}$ and its dependence cone $\Theta_{\mathcal{I}}^{*}$ is pointed (see Fig. 3.1).   ■ 3.1.7

3.1.1   Integral *vs.* Affine Recurrences

Because of the generality of our definition of an integral index mapping, it is easy to show that affine index mappings constitute a particular type of integral index mappings.

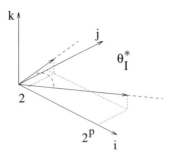

Fig. 3.1. Pointed dependence cone.

Let us consider an affine index mapping $\mathcal{I}(z) = Az + b$, with $A \in \mathbf{Z}^{n \times n}$ and $b \in \mathbf{Z}^n$. Then $\mathcal{I}$ can be expressed as an integral index mapping where, for $j = 1, \ldots, n$:

- each coefficient g_j is defined by the vector expression $g_j(z) = (A - \mathbf{I}_n)_j \cdot z + b_j$, with $(A - \mathbf{I}_n)_j$ denoting the jth row of the matrix $A - \mathbf{I}_n$, and b_j the jth component of b; and

- each generator e_j is the jth vector of the standard basis of $\mathbf{Z}^n$.

That is the $\mathcal{I}$ can be expressed as follows::

$$
\begin{aligned}
\mathcal{I}(z) &= Az + b \\
&= z + (A - \mathbf{I}_n)z + b \\
&= z + \sum_{j=1}^{n} g_j(z)e_j.
\end{aligned}
$$

Trivially, all uniform index mappings are atomic integral. In fact, a uniform index mapping has a generic form $\mathcal{I}(z) = z + b$, which can be seen as an integral form with a single generator b and, as coefficient, the constant function $g(z) = 1$.

Given the above relations, a taxonomy of the different classes of recurrences can be defined as illustrated in Fig. 3.2, in which we follow the custom of naming recurrence equations after the (most general) type of their data dependencies. In particular, affine recurrences (AREs) constitute a proper subclass of integral recurrences (IREs), while uniform recurrences (UREs) are in particular atomic integral (AIREs).

Therefore, integral recurrence equations can be used as more general forms of specification than those obtained with affine recurrences only. For

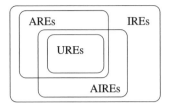

Fig. 3.2. Inclusions among the classes of uniform (UREs), affine (AREs), atomic integral (AIREs) and integral (IREs) recurrence equations.

specification in the intersection of the two classes, the convenience of one formalism over the other needs to be evaluated case by case. In particular, a trade-off between ease of expression and complexity of the required regularisation should be considered. We will return to this point in the discussion at the end of the chapter.

3.2 Regularisation

The regularisation of an integral data dependence consists of two main steps:

- the substitution of a generic integral data dependence by a set of atomic integral data dependencies; and

- the substitution of each atomic integral data dependence by a set of uniform data dependencies.

Routing directions are chosen among the generators of the data dependence mapping, which guarantee the definition of injective index mappings, hence no overloaded data dependence graphs are generated. The major difference with respect to regularisation techniques for the affine case is the necessity of using control variables to reconcile the preservation of convexity with the application of non-affine transformations.

3.2.1 Regularisation Directions

Let $\mathcal{DD} = (D, U, V, \mathcal{I})$ be an integral data dependence, with pointed dependence cone $\Theta_{\mathcal{I}}^*$. We intend to use a set of non-null generators of $\mathcal{I}$ as regularisation direction vectors. To do so, we want to express $\mathcal{I}$ so that the generators of $\mathcal{I}$ form a pointed cone C containing $\Theta_{\mathcal{I}}^*$. Such a representation is particularly convenient because:

- it allows us to formulate conditions for the existence and preservation of affine scheduling through regularisation;

- it simplifies the definition of regularisation techniques, as the coefficients define non-negative integer functions on D.

Such a form always exists, as can be proved by using a result due to Quinton and Van Dongen [QuVa89]. The result is contained in the following proposition:

Proposition 3.2.1 [QuVa89] Let C be a pointed polyhedral convex cone of full dimension in $\mathbf{Q}^n$. There exists a pointed polyhedral convex cone C' such that: C' contains C and its extremal rays constitute a unimodular basis of $\mathbf{Z}^n$.

PROOF: The proof is based on the two following properties of convex polyhedral cones:

- if C and C' are convex polyhedral cones such that $C \subseteq C'$, then their dual cones $\hat{C}$ and $\hat{C}'$ are such that $\hat{C}' \subseteq \hat{C}$;

- if C is an n-dimensional cone in $\mathbf{Q}^n$, with n extremal rays $r_1, \ldots, r_n$ forming the columns of a matrix Q, then if $\hat{C}$ is an n-dimensional cone in $\hat{\mathbf{Q}}^n$, with n extremal rays forming the columns of the matrix $\hat{Q} = -(Q^{-1})^t$.

Suppose C has m extremal rays, with $m \geq n$ (m cannot be less than n as C is assumed of full dimension in $\mathbf{Q}^n$). Let $\hat{C}$ be its dual cone and let $\hat{R}$ be the matrix having as columns the m extremal rays of $\hat{C}$. Choose a sub-cone $\hat{C}'$ of $\hat{C}$ such that $\hat{C}'$ has exactly n extremal rays which form a unimodular basis. Such rays are the columns of any $n \times n$ unimodular matrix $\hat{R}'$, which satisfies the matrix equation:

$$\hat{R} \cdot P = \hat{R}'$$

with P a non-negative $m \times n$ rational matrix. Then the dual cone C' of $\hat{C}'$ is the cone we are looking for, with extremal rays the columns of the matrix $-(\hat{R}'^{-1})^t$.   ■ 3.2.1

We illustrate the result in the following example.

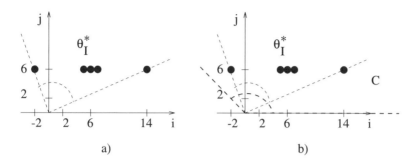

a)   b)

Fig. 3.3. a) Data dependence cone $\Theta_{\mathcal{I}}^*$; b) Pointed cone C containing the cone $\Theta_{\mathcal{I}}^*$.

Example 3.2.2   Consider the data dependence $\mathcal{DD} = (D, U, V, \mathcal{I})$ with domain $D = \{(i, j) \mid -2 \leq i \leq 2, 1 \leq j \leq 2\}$ and index mapping $\mathcal{I}(i, j) = (i, j) + 6(-1, -1) + i^3(-1, 0)$. The dependence mapping is $\Theta_{\mathcal{I}}(i, j) = 6(1, 1) + i^3(1, 0)$ and $\Theta_{\mathcal{I}}^*$ is illustrated in Fig. 3.3 a). The dependence cone is pointed. However, the coefficient i^3 does not define a non-negative integer function over D.

Consider the pointed cone $C = cone(\{(-1, 1), (1, 0)\})$. C contains $\Theta_{\mathcal{I}}^*$ (see Fig. 3.3 b)) and its generators constitute a unimodular basis (i.e., they integrally span $\mathbf{Z}^2$). The correspondence between the generators of $\Theta_{\mathcal{I}}$ and those of C is the following:

$$\begin{aligned}
(1, 1) &= 1(-1, 1) + 2(1, 0) \\
(1, 0) &= 0(-1, 1) + 1(1, 0).
\end{aligned}$$

Therefore we can rewrite $\Theta_{\mathcal{I}}$ and $\mathcal{I}$ as follows:

$$\begin{aligned}
\Theta_{\mathcal{I}}(i, j) &= 6[1(-1, 1) + 2(1, 0)] + i^3[0(-1, 1) + 1(1, 0)] \\
&= 6(-1, 1) + (12 + i^3)(1, 0) \\
\mathcal{I}(i, j) &= (i, j) + 6(1, -1) + (12 + i^3)(-1, 0).
\end{aligned}$$

With this rewriting, the coefficients of $\mathcal{I}$ are non-negative over D.   ■3.2.2

Note that if $\Theta_{\mathcal{I}}^*$ is pointed and the coefficients of $\Theta_{\mathcal{I}}$ are non-negative integer functions on D, the cone C defined by the generators of $\Theta_{\mathcal{I}}$ contains the dependence cone $\Theta_{\mathcal{I}}^*$. However, C is not guaranteed to be pointed, as shown in the following example.

Fig. 3.4. Dependence cones: a) $\Theta_{\mathcal{I}}^*$; b) C.

Example 3.2.3   Consider the data dependence $\mathcal{DD} = (D, U, V, \mathcal{I})$ with domain $D = \{(i, j) \mid 1 \leq i, j \leq m\}$, where $m \in \mathbf{N}$, and index mapping $\mathcal{I}(i, j) = (i, j) + 6(1, 0) + 4(-1, 0)$. The dependence cone $\Theta_{\mathcal{I}}^*$ is pointed. However, $C = cone(\{(1, 0), (-1, 0)\})$ contains a line (see Fig. 3.4). Indeed, a straightforward rewriting of $\mathcal{I}$ exists such that $\Theta_{\mathcal{I}}^*$ and C coincide. The rewriting is $\mathcal{I}(i, j) = (i, j) + (6 - 4)(1, 0) = (i, j) + 2(1, 0)$.   ■ 3.2.3

   In the following, we always assume that given an integral data dependence $\mathcal{DD} = (D, U, V, \mathcal{I})$, the generators of $\mathcal{I}$ are non-null vectors, its coefficients define non-negative integer functions on D, and the cone C defined by the generators of $\Theta_{\mathcal{I}}$ is pointed. We will also refer to the cone C as the *embedding dependence cone*.

3.2.2   Injectivity of an Atomic Integral Index Mapping

From the discussion in Section 2.3, we know that a data broadcast corresponds to a non-injective index mapping. A sufficient condition for the injectivity of an atomic integral index mapping $\mathcal{I}$ on a domain D can be established by considering the relation between the generator of $\mathcal{I}$ and the direction $lin(D)$ of the domain D (for a definition of $lin(D)$, see Appendix D). The result is contained in the following proposition:

Proposition 3.2.4   Consider an atomic integral index mapping $\mathcal{I}$ and a domain $D \subseteq \mathbf{Z}^n$, such that for all $z \in D$, $\mathcal{I}(z) = z + g(z)d$ and $g(z) \geq 0$. If $d \notin lin(D)$ then $\mathcal{I}$ is injective over D.

PROOF: From linear algebra, if $z, z' \in D$ then $z - z' \in lin(D)$. Assume that there exist $z, z' \in D$, such that $z \neq z'$ and $\mathcal{I}(z) = \mathcal{I}(z')$. We want to prove that this assumption always implies a contradiction with respect to the hypotheses of the proposition and therefore for all $z, z' \in D$, $z \neq z'$ implies $\mathcal{I}(z) \neq \mathcal{I}(z')$. There are only two

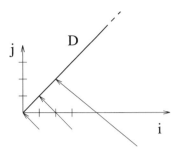

Fig. 3.5. Injectivity.

possibilities, both leading to a contradiction. If $g(z) = g(z') = c$, then $\mathcal{I}(z) = \mathcal{I}(z')$ implies $z + cd = z' + cd$, i.e., $z = z'$. Otherwise, if $g(z) \neq g(z')$, then $g(z') - g(z) = c \neq 0$ and $\mathcal{I}(z) = \mathcal{I}(z')$ implies $z - z' = cd$, i.e., $d \in lin(D)$.   ∎ 3.2.4

Example 3.2.5   Consider a data dependence $\mathcal{DD} = (D, U, V, \mathcal{I})$ with domain $D = \{(i, j) \mid i \geq 0, i = j\}$ and integral index mapping $\mathcal{I}(i, j) = (i, j) + 2^j(1, -1)$. The direction of D is $lin(D) = \langle (1, 1) \rangle$ and $d = (1, -1) \notin lin(D)$. Hence, according to Proposition 3.2.4, $\mathcal{I}$ is injective over D. Some of the corresponding data dependence vectors are sketched in Fig. 3.5.
                                                                        ∎ 3.2.5

Note that the condition is only sufficient and there may exist atomic index mappings which are injective regardless of the geometric relation between their generators and the domain. This is illustrated in the following example.

Example 3.2.6   Consider the atomic index mapping $\mathcal{I}(i, j) = (i, j) + j(1, 0)$ and the domain $D = \{(i, j) \mid 1 \leq i, j \leq p\}$, for some $p > 1$. As D is of full dimension in $\mathbf{Z}^2$, then $d \in lin(D)$. However, $\mathcal{I}$ is injective on D. In particular $\mathcal{I}$ defines an injective mapping on $\mathbf{Z}^2$.   ∎ 3.2.6

Given an atomic integral index mapping which is injective according to Proposition 3.2.4, we want to define a corresponding inverse mapping. Inverse mappings will be needed in the definition of the decomposition techniques for integral index mappings in Section 3.2.5.

An inverse of an atomic integral index mapping which satisfies the condition of Proposition 3.2.4, can be obtained in a systematic way, based only on geometric properties. This definition is given in Proposition 3.2.7 below, and is based on the following:

- the choice of a hyperplane $[\pi : \theta]$, with $\pi \neq 0$, containing D. By definition, π has to be a vector orthogonal to all vectors in D. Hence π has to be chosen in the space $D^{\perp}$ (the orthogonal complement of D - see Appendix D for a definition). We also require that π is in the space generated by D and d. This choice will allow us to define a measure of the distance from D of a point in the direction of d. Formally, this choice is expressed by the condition $\pi \in lin(P) \cap D^{\perp}$ of the proposition, where P is the polyhedron generated by D and d (see also Appendix C). The effect of this choice is illustrated in Fig. 3.6 a) in $\mathbf{Z}^2$, for a 1-dimensional domain D. Fig. 3.6 b) illustrates a (non-admissible) choice of π outside the polyhedron P.

- the scalar product $\eta = \pi \cdot d$, which represents the projection of d along the direction of π. The integer η may be seen as establishing a "distance" between parallel hyperplanes with normal vector π intersecting the polyhedron P (see Fig. 3.6 c)).

- a linear transformation l, defined so that for all $z \in D$, $l(z) = g(z)$. For each point $z \in D$, $l(z)$ defines the "distance" (as a multiple of η) between the parallel hyperplanes $[\pi : \theta]$ and $[\pi : \theta + g(z)\eta]$ (see Fig. 3.6 d)).

Proposition 3.2.7   Consider an atomic integral index mapping $\mathcal{I}$ and a domain $D \subseteq \mathbf{Z}^n$, such that for all $z \in D$, $\mathcal{I}(z) = z + g(z)d$ and $g(z) \geq 0$. Let $d \notin lin(D)$ and P the convex polyhedron generated by D and d. Consider $\pi \in lin(P) \cap D^{\perp}$, with $\pi \neq 0$, and the hyperplane $[\pi : \theta]$ containing the domain D.

Then the mapping $\mathcal{I}'(z) = z + l(z)(-d)$, where $l(z) = (\pi \cdot z - \theta)/\eta$ and $\eta = \pi \cdot d$, defines an inverse of $\mathcal{I}$ over D.

PROOF: By definition, for all $z \in D$, $\pi \cdot z = \theta$. Let $z \in D$, then:

$$l(\mathcal{I}(z)) = l(z + g(z)d) = (\pi \cdot (z + g(z)d) - \theta)/\eta$$
$$= (\pi \cdot z + g(z)\pi \cdot d - \theta)/\eta = (\theta + g(z)\eta - \theta)/\eta = g(z).$$

Therefore, for all $z \in D$,

$$\mathcal{I}' \circ \mathcal{I}(z) = \mathcal{I}'(\mathcal{I}(z)) = \mathcal{I}(z) + l(\mathcal{I}(z))(-d) = z + g(z)d - g(z)d = z.$$

$$\blacksquare \ 3.2.7$$

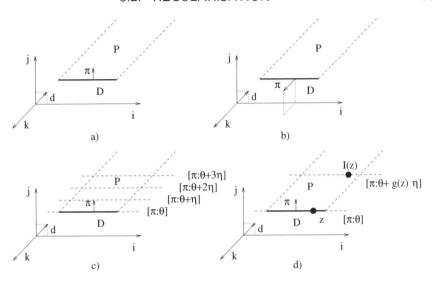

Fig. 3.6. Inverse of an injective atomic integral index mapping: a) admissible choice of π; b) non-admissible choice of π; c) intuitive meaning of η; d) intuitive meaning of l.

Example 3.2.8   Let us consider the atomic integral index mapping of Example 3.2.5, and apply Proposition 3.2.7. Let $\pi = (1, -1)$, $\theta = 0$ and $\eta = \pi \cdot d = (1, -1) \cdot (1, -1) = 2$. Then $l(i, j) = (i - j)/2$ and $\mathcal{I}'(i, j) = ((i + j)/2, (i + j)/2)$. For instance, $\mathcal{I}' \circ \mathcal{I}(3, 3) = \mathcal{I}'(11, -5) = (6/2, 6/2) = (3, 3)$.
                                                                        ■ 3.2.8

Proposition 3.2.7 provides a way of defining an inverse for an atomic index mapping (which satisfies the conditions of the proposition) that can be easily mechanised. It may be the case that an inverse for the mapping can be provided by the algorithm designer by other means. This happens, in particular, for injective linear mappings, as shown in the following example.

Example 3.2.9   Consider the index mapping $\mathcal{I}(i, j) = (i + j, j)$ of Example 3.2.6. $\mathcal{I}$ defines an injective linear mapping in $\mathbf{Z}^n$. In its matrix form, $\mathcal{I}$ can be expresses as:

$$\mathcal{I}\begin{pmatrix} i \\ j \end{pmatrix} = \begin{bmatrix} 1 & 1 \\ 0 & 1 \end{bmatrix} \cdot \begin{pmatrix} i \\ j \end{pmatrix}$$

An inverse for $\mathcal{I}$ can be determine by computing the inverse of the matrix

$$A = \begin{bmatrix} 1 & 1 \\ 0 & 1 \end{bmatrix},$$

that is the matrix

$$A^{-1} = \begin{bmatrix} 1 & -1 \\ 0 & 1 \end{bmatrix}.$$

Therefore, the mapping $\mathcal{I}'(i, j, k) = (i - j, j)$ defines an inverse of $\mathcal{I}$.

∎ 3.2.9

From the above discussion, it follows that we may consider two ways of deciding whether an atomic integral index mapping is injective on a certain domain, as well as defining a corresponding inverse: automatically, by considering geometric properties only, or in an *ad hoc* fashion by direct intervention of the algorithm designer. The two approaches are complementary and should both be accounted for by the method. In Section 3.2.5 we will take advantage of both for the definition of decomposition techniques.

3.2.3   Uniformisation

Given an atomic integral data dependence $\mathcal{DD}$, uniformisation defines a substitution of $\mathcal{DD}$ with a uniform routing system defined by a set of uniform recurrence equations.

Let $\mathcal{I}$ denote the atomic integral index mapping of $\mathcal{DD}$ and g its coefficient. Uniformisation can be applied only if g admits an upper bound on the domain D, i.e., there exists $m \in \mathbf{N}$ such that for all $z \in D$, $g(z) \leq m$. Technically, it is this condition which allows us to define and initialise the control variables which control the uniform routing of the data. Although this is a restriction to the application of the technique, it is justifiable in terms of requirements for a realistic implementation. In fact, if g is not bounded over D, the Euclidean distance between pairs of data dependent points increases without a bound, and so does the order of the routing paths which are defined between the two points. Because of the relation between data dependence graph and data flow graph, this fact results in unbounded resources (either memory or number of processing elements and connections) of any regular array implementation of the specification.

The following propositions define two uniformisation techniques with the guarantee that no problem size dependent overloading of the data dependence graph is generated. The difference between the two techniques is the geometric relation between the direction vector d and the domain D of the data dependence. In particular, if $d \notin lin(D)$, because of Proposition 3.2.4, $\mathcal{I}$ is injective and d can be used as the direction for pipelining the data. Otherwise, new pipelining directions need to be found outside $lin(D)$, which guarantee the injectivity of the corresponding index mappings.

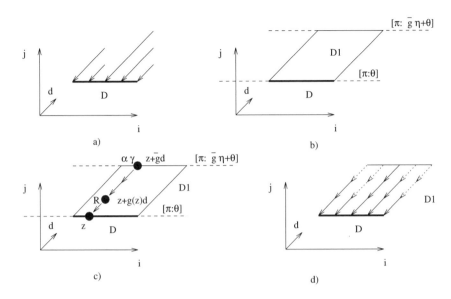

Fig. 3.7. Uniformisation 1.

We use control variables in the definition of the routing schemes. The need for control variables comes from the contrasting necessity of obtaining polyhedral convex regions even when non-affine mappings are applied. As, in general, $\mathcal{I}$ is not linear, the image of D under $\mathcal{I}$ is not a polyhedral convex set. However, polyhedral convex routing domains are enforced, which embed the actual routing paths of the data. The points needed for routing constitute non-convex sub-sets of such domains. The identification of the required points is achieved via conditional expressions based on the integer coefficient g.

The content of Proposition 3.2.10 is illustrated in Fig. 3.7, for a 2-dimensional case. The aim is to transform the data dependence graph in Fig. 3.7 a) into that of Fig. 3.7 d). A new domain D_1 is defined from D and the maximum value $\bar{g}$ that the coefficient g assumes on D (see Fig. 3.7 b)). The choice of π and the hyperplane $[\pi : \theta]$ containing D, already discussed in Section 3.2.2, guarantees that no data broadcast is generated. In Fig. 3.7 d) solid arrows correspond to the routing paths which substitute the data dependence vectors of Fig. 3.7 a). The dashed arrows result because of the uniformisation on the enlarged domains and do not contribute to the actual routing. The computations on the routing paths are identified by the evaluation of control variables α and γ. The definition of α and γ is based on the

coefficient g of the index mapping, and their use is illustrated in Fig. 3.7 c).
In the figure, three points of the domain D_1 are emphasised: point z needs
to receive the data from $z + g(z)d$; variables α and γ are initialised at $z + \bar{g}d$.
Note that the three points lie on the line $z + ld$, for $l \in \mathbf{R}$. At $z + \bar{g}d$, variable
α is initialised to the value $g(z)$, and γ to $\bar{g}$. Hence, initially, $\gamma \geq \alpha$. The
control data α and γ are then pipelined according to the direction of $-d$. At
each step, the value of γ is decreased by 1. Also, at each step the values of α
and γ are compared, and when they become equal (i.e., both equal to $g(z)$),
the data is collected by a routing variable R and subsequently pipelined to
z along $-d$.

Proposition 3.2.10 *[Uniformisation 1]* Let $\mathcal{DD} = (D, U, V, \mathcal{I})$ be an atomic integral data dependence, with $\mathcal{I}(z) = z + g(z)d$. Let g be non-negative
and bounded over D, and $\bar{g}$ the least upper bound of g. Let $d \notin lin(D)$ and P
be the convex polyhedron generated by D and d. Consider $\pi \in lin(P) \cap D^{\perp}$,
with $\pi \neq 0$, and the hyperplane $[\pi : \theta]$ containing the domain D.
    Then $\mathcal{DD}$ can be substituted by the uniform data dependence

$$\mathcal{DD}' \;=\; (D, U, R, \mathcal{I}_0)$$

and the system of equations:

$$
\begin{aligned}
\mathbf{E}_1 &= (D_1, R, (R, V, \alpha, \gamma), f, (\mathcal{I}_1, \mathcal{I}_0, \mathcal{I}_0, \mathcal{I}_0)) \\
\mathbf{E}_2 &= (D_{1,1}, \alpha, \alpha, id, \mathcal{I}_1) \\
\mathbf{E}_3 &= (D_{1,2}, \alpha, in_\alpha) \\
\mathbf{E}_4 &= (D_{1,1}, \gamma, \gamma, dec, \mathcal{I}_1) \\
\mathbf{E}_5 &= (D_{1,2}, \gamma, in_\gamma)
\end{aligned}
$$

where:

- the index mappings are:

$$
\begin{aligned}
\mathcal{I}_0(z) &= z \\
\mathcal{I}_1(z) &= z + d
\end{aligned}
$$

- R, α and γ are new variables;

- the applied functions are:

$$in_\alpha(z) \;=\; g(z - \bar{g}d)$$

$$in_\gamma(z) \; = \; \bar{g}$$
$$id(a) \; = \; a$$
$$dec(a) \; = \; a - 1$$
$$f(a, b, c, d) \; = \; \begin{cases} a & c \neq d \\ b & \text{otherwise} \end{cases}$$

- the new domains are:

$$D_1 \; = \; \{z + ld \mid z \in D, 0 \leq l \leq \bar{g}\}$$
$$D_{1,1} \; = \; \{z + ld \mid z \in D, 0 \leq l < \bar{g}\}$$
$$D_{1,2} \; = \; \{z + \bar{g}d \mid z \in D\}$$

PROOF: For all $z \in D$, let $segm(z) = \{z + ld \mid 0 \leq l \leq \bar{g}\}$. By definition, for all $z' \in segm(z)$, $\alpha(z') = g(z)$. In fact, for $z \in D$,

$$\alpha(z) = \alpha(z + d)$$
$$= \; \ldots = \alpha(z + \bar{g}d) = g(z + \bar{g}d - \bar{g}d) = g(z).$$

Also for all $z' = z + ld \in segm(z)$, $\gamma(z') = l$. In fact, for $z \in D$,

$$\gamma(z) = \gamma(z + d) - 1$$
$$= \; \ldots = \gamma(z + \bar{g}d) - \bar{g} = \bar{g} - \bar{g} = 0.$$

Therefore, for all $z' = z + ld \in segm(z)$, $\alpha(z') = \gamma(z')$ if and only if $l = g(z)$.

Hence, for all $z \in D$,

$$U(z) = R(\mathcal{I}_0(z)) = R(\mathcal{I}_1 \circ \mathcal{I}_0(z))$$
$$= \; \ldots = R(\mathcal{I}_1^{g(z)} \circ \mathcal{I}_0(z)) = V(\mathcal{I}_0 \circ \mathcal{I}_1^{g(z)} \circ \mathcal{I}_0(z))$$
$$= \; V(z + g(z)d) = V(\mathcal{I}(z)).$$

■ 3.2.10

Corollary 3.2.11   Let $\mathcal{DD} = (D, U, V, \mathcal{I})$ be an atomic integral data dependence, with $\mathcal{I}(z) = z + g(z)d$. Let g be non-negative and bounded over D, and $\bar{g}$ the least upper bound of g.

If $\bar{g} = 0$, then $\mathcal{DD}$ can be substituted by the uniform data dependence

$$\mathcal{DD}' \; = \; (D, U, V, \mathcal{I}_0)$$

where $\mathcal{I}_0(z) = z$.

■ 3.2.11

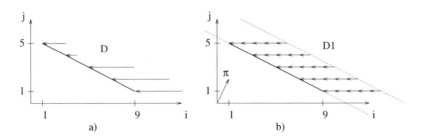

Fig. 3.8. Uniformisation: a) domain D and corresponding data dependence vectors; b) domain D_1 and data dependence vectors after uniformisation.

Example 3.2.12   Consider the atomic integral data dependence $\mathcal{DD} = (D, U, V, \mathcal{I})$ with index mapping $\mathcal{I}(i, j) = (i, j) + g(i, j)(1, 0)$, where $g(i, j) = (i^2 + j^2)mod\ 6$, and domain $D = \{(i, j) \mid 1 \leq j \leq 5, i = 11 - 2j\}$. $lin(D) = \langle(2, -1)\rangle$ and $d = (1, 0) \notin lin(D)$. $\mathcal{DD}$ is illustrated in Fig. 3.8 a).

The space generated by D and d is the whole $\mathbf{Z}^2$, and π can be any vector in $D^\perp$. For instance, the vector $\pi = (1, 2)$ (which is orthogonal to the generator $(2, -1)$ of $lin(D)$) satisfies the conditions of Proposition 3.2.10, and $[(1, 2) : 11]$ is a hyperplane containing D. Also, for all $(i, j) \in D$, $g(i, j) \leq 5$. If we apply Proposition 3.2.10 we obtain the new data dependence

$$\mathcal{DD}' \;=\; (D, U, R, \mathcal{I}_0)$$

and the system of equations:

$$
\begin{aligned}
\mathbf{E}_1 &= (D_1, R, (R, V, \alpha, \gamma), f, (\mathcal{I}_1, \mathcal{I}_0, \mathcal{I}_0, \mathcal{I}_0)) \\
\mathbf{E}_2 &= (D_{11}, \alpha, \alpha, id, \mathcal{I}_1) \\
\mathbf{E}_3 &= (D_{12}, \alpha, in_\alpha) \\
\mathbf{E}_4 &= (D_{11}, \gamma, \gamma, dec, \mathcal{I}_1) \\
\mathbf{E}_5 &= (D_{12}, \gamma, in_\gamma)
\end{aligned}
$$

where:

- the index mappings are:

$$
\begin{aligned}
\mathcal{I}_0(i, j) &= (i, j) \\
\mathcal{I}_1(i, j) &= (i, j) + (1, 0)
\end{aligned}
$$

- the applied functions are:

$$in_\alpha(i, j) \;=\; g(i - 5, j)$$

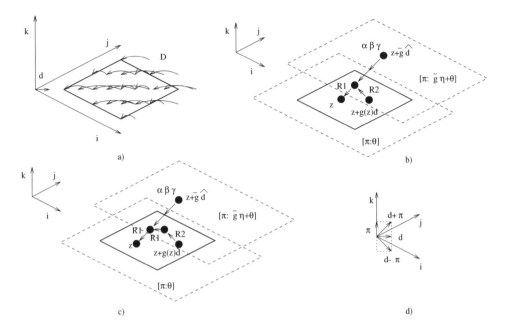

Fig. 3.9. Uniformisation 2.

$$in_\gamma(z) = 5$$
$$id(a) = a$$
$$dec(a) = a - 1$$
$$f(a, b, c, d) = \begin{cases} a & c \neq d \\ b & \text{otherwise} \end{cases}$$

- the new domains are:

$$D_1 = \{(i, j) \mid 1 \leq j \leq 5, 11 - 2j \leq i \leq 16 - 2j\}$$
$$D_{1,1} = \{(i, j) \in D_1 \mid i + 2j < 16\}$$
$$D_{1,2} = \{(i, j) \in D_1 \mid i + 2j = 16\}.$$

The resulting data dependence graph is illustrated in Fig. 3.8 b). The correct routing of the data is achieved by evaluating the control variables α and γ. For simplicity, in the figure, the propagation of the control variables has been omitted.   ■ 3.2.12

The second uniformisation technique applies when the generator d of the index mapping $\mathcal{I}$ cannot be used as a regularisation direction. Fig. 3.9

illustrates the result in a 3-dimensional space. In this case we assume that d is contained in $lin(D)$ (in Fig. 3.9 a), $lin(D)$ is the space generated by $(1,0,0)$ and $(0,1,0)$). A vector π is chosen in $D^\perp$ (in the figure, $\pi = (0,0,1)$). Three regularisation vectors are involved: d itself, together with $\hat{d} = d + \pi$ and $\check{d} = d - \pi$ (see Fig. 3.9 d)). Also, three control variables, α, β and γ, are used. The effect of uniformisation is illustrated in Fig. 3.9 b) and c) for $g(z)$ even and odd, respectively. The data dependence vector between z and $z + g(z)d$ is replaced by a routing path of order $g(z)$. This path is the result of:

- a sub-path of order $\lfloor g(z)/2 \rfloor$ according to the direction of $\check{d}$, where $\lfloor g(z)/2 \rfloor$ denotes the floor function applied to $g(z)/2$, which returns the largest integer smaller than or equal to $g(z)/2$. On this sub-path the data is pipelined by a routing variable R^2 (R2 in the figure);

- a sub-path of order $g(z) \bmod 2$ according to the direction of d, where mod denotes the modulo function. On this sub-path the data is pipelined by a routing variable R^1 (R1 in the figure). This is necessary only if $g(z)$ is odd, as in Fig. 3.9 c);

- a sub-path of order $\lfloor g(z)/2 \rfloor$ according to the direction of $\hat{d}$. On this sub-path the data is also pipelined by the routing variable R^1.

The control variables α, β and γ are initialised at $z + \bar{g}\hat{d}$, where $\bar{g} = \lfloor m/2 \rfloor$ and m is the least upper bound of g on D. Variable α is initialised to $\lfloor g(z)/2 \rfloor$, γ to $\bar{g}$, and β to $g(z) \bmod 2$. The values of α, β and γ are pipelined according to the direction of $-\hat{d}$, with the value of γ decreased by one at each step. Also, the values of α and γ are compared at each step; when they become equal (i.e., both equal to $\lfloor g(z)/2 \rfloor$), the value of β is considered to determine whether $g(z)$ is either even or odd, and the data is transferred accordingly from R^2 to R^1.

Proposition 3.2.13 *[Uniformisation 2]* Let $\mathcal{DD} = (D, U, V, \mathcal{I})$ be an atomic finitely generated dynamic data dependence, with $\mathcal{I}(z) = z + g(z)d$. Let g be non-negative and bounded over D, m the least upper bound of g and $\bar{g} = \lfloor m/2 \rfloor$. Let $d \in lin(D)$ and $dim(D) < n$. Consider $\pi \in D^\perp$, with $\pi \neq 0$, the hyperplane $[\pi : \theta]$ containing the domain D, and let $\hat{d} = d + \pi$ and $\check{d} = d - \pi$.

Then $\mathcal{DD}$ can be substituted by the uniform data dependence

$$\mathcal{DD}' \quad = \quad (D, U, R^1, \mathcal{I}_0)$$

and the system of equations:

$$
\begin{aligned}
\mathbf{E}_1 &= (D_1, R^1, (R^1, R^2, R^2, \alpha, \beta, \gamma), f, (\mathcal{I}_1, \mathcal{I}_{2,0}, \mathcal{I}_{2,1}, \mathcal{I}_0, \mathcal{I}_0, \mathcal{I}_0)) \\
\mathbf{E}_2 &= (D_{2,1}, R^2, R^2, id, \mathcal{I}_3) \\
\mathbf{E}_3 &= (D_{2,2}, R^2, V, id, \mathcal{I}_0) \\
\mathbf{E}_4 &= (D_{1,1}, \alpha, \alpha, id, \mathcal{I}_1) \\
\mathbf{E}_5 &= (D_{1,2}, \alpha, in_\alpha) \\
\mathbf{E}_6 &= (D_{1,1}, \beta, \beta, id, \mathcal{I}_1) \\
\mathbf{E}_7 &= (D_{1,2}, \beta, in_\beta) \\
\mathbf{E}_8 &= (D_{1,1}, \gamma, \gamma, dec, \mathcal{I}_1) \\
\mathbf{E}_9 &= (D_{1,2}, \gamma, in_\gamma)
\end{aligned}
$$

where:

- the index mappings are:

$$
\begin{aligned}
\mathcal{I}_0(z) &= z \\
\mathcal{I}_1(z) &= z + \hat{d} \\
\mathcal{I}_{2,0}(z) &= z \\
\mathcal{I}_{2,1}(z) &= z + d \\
\mathcal{I}_3(z) &= z + \check{d}
\end{aligned}
$$

- R^1, R^2, α, β and γ are new variables;

- the applied functions are:

$$
\begin{aligned}
in_\alpha(z) &= \lfloor g(z - \bar{g}\hat{d})/2 \rfloor \\
in_\beta(z) &= g(z - \bar{g}\hat{d}) \bmod 2 \\
in_\gamma(z) &= \bar{g} \\
id(a) &= a \\
dec(a) &= a - 1 \\
f(a, b, c, d, e, f) &= \begin{cases} a & d \neq f \\ b & d = f, e = 0 \\ c & d = f, e = 1 \end{cases}
\end{aligned}
$$

- the new domains are:

$$
D_1 = \{z + l\hat{d} \mid z \in D, 0 \leq l \leq \bar{g}\}
$$

$$
\begin{aligned}
D_{1,1} &= \{z + l\hat{d} \mid z \in D, 0 \le l < \bar{g}\} \\
D_{1,2} &= \{z + \bar{g}\hat{d} \mid z \in D\} \\
D_2 &= \{z + l_1 d + l_2 \check{d} \mid z \in D_1, 0 \le l_1 \le 1, 0 \le l_2 \le \bar{g}\} \cap \\
&\quad \{z \in \mathbf{Z}^n \mid \pi \cdot z \ge \theta\} \\
D_{2,1} &= \{z \in D_2 \mid \pi \cdot z > \theta\} \\
D_{2,2} &= \{z \in D_2 \mid \pi \cdot z = \theta\}.
\end{aligned}
$$

PROOF: For all $z \in D$, let $segm(z) = \{z + ld \mid 0 \le l \le \bar{g}\}$. By definition, for all $z' \in segm(z)$, $\alpha(z') = \lfloor g(z)/2 \rfloor$. In fact, for $z \in D$,

$$
\begin{aligned}
\alpha(z) &= \alpha(z + \hat{d}) \\
&= \ldots = \alpha(z + \bar{g}\hat{d}) = \lfloor g(z + \bar{g}\hat{d} - \bar{g}\hat{d})/2 \rfloor = \lfloor g(z)/2 \rfloor.
\end{aligned}
$$

Also, for all $z' \in segm(z)$, $\beta(z') = g(z) \bmod 2$. In fact, for $z \in D$,

$$
\begin{aligned}
\beta(z) &= \beta(z + \hat{d}) \\
&= \ldots = \beta(z + \bar{g}\hat{d}) = g(z + \bar{g}\hat{d} - \bar{g}\hat{d}) \bmod 2 = g(z) \bmod 2.
\end{aligned}
$$

Finally, for all $z' = z + l\hat{d} \in segm(z)$, $\gamma(z') = l$. In fact, for $z \in D$,

$$
\begin{aligned}
\gamma(z) &= \gamma(z + \hat{d}) - 1 \\
&= \ldots = \gamma(z + \bar{g}\hat{d}) - \bar{g} = \bar{g} - \bar{g} = 0.
\end{aligned}
$$

Therefore, for all $z' = z + ld \in segm(z)$, $\alpha(z') = \gamma(z')$ if and only if $l = \lfloor g(z)/2 \rfloor$.

We observe that, for all $c \in \mathbf{Z}$,

$$
\begin{aligned}
\lfloor c/2 \rfloor &= (c - c \bmod 2)/2 \\
c &= 2\lfloor c/2 \rfloor + c \bmod 2.
\end{aligned}
$$

Hence, for all $z \in D$,

$$
\begin{aligned}
U(z) &= R^1(\mathcal{I}_0(z)) = R^1(\mathcal{I}_1 \circ \mathcal{I}_0(z)) \\
&= \ldots = R^1(\mathcal{I}_1^{\lfloor g(z)/2 \rfloor} \circ \mathcal{I}_0(z)) \\
&= R^2(\mathcal{I}_{2,g(z) \bmod 2} \circ \mathcal{I}_1^{\lfloor g(z)/2 \rfloor} \circ \mathcal{I}_0(z)) \\
&= R^2(\mathcal{I}_3 \circ \mathcal{I}_{2,g(z) \bmod 2} \circ \mathcal{I}_1^{\lfloor g(z)/2 \rfloor} \circ \mathcal{I}_0(z)) \\
&= \ldots = R^2(\mathcal{I}_3^{\lfloor g(z)/2 \rfloor} \circ \mathcal{I}_{2,g(z) \bmod 2} \circ \mathcal{I}_1^{\lfloor g(z)/2 \rfloor} \circ \mathcal{I}_0(z))
\end{aligned}
$$

$$
\begin{aligned}
&= V(\mathcal{I}_0 \circ \mathcal{I}_3^{\lfloor g(z)/2 \rfloor} \circ \mathcal{I}_{2,g(z) \bmod 2} \circ \mathcal{I}_1^{\lfloor g(z)/2 \rfloor} \circ \mathcal{I}_0(z)) \\
&= V(z + \lfloor g(z)/2 \rfloor \hat{d} + (g(z) \bmod 2)d + \lfloor g(z)/2 \rfloor \check{d}) \\
&= V(z + 2\lfloor g(z)/2 \rfloor d + (g(z) \bmod 2)d) \\
&= V(z + (2\lfloor g(z)/2 \rfloor + g(z) \bmod 2)d) \\
&= V(z + g(z)d) = V(\mathcal{I}(z)).
\end{aligned}
$$

∎ 3.2.13

Because of the condition $dim(D) < n$, the application of Proposition 3.2.13 may require an increase in the number of dimensions of the lattice space and to reindex the data dependence accordingly. Such a reindexing is a particular case of normalisation as described in Section 2.2.2 and is realised by "padding" the index expressions with extra indices. Note that this normalisation should not alter the dimensionality of the domains (it simply produces an embedding of the domains in a space of higher dimension).

Corollary 3.2.14 Let $\mathcal{DD} = (D, U, V, \mathcal{I})$ be an atomic integral data dependence, with $\mathcal{I}(z) = z + g(z)d$. Let g be non-negative and bounded over D, m the least upper bound of g, and $\bar{g} = \lfloor m/2 \rfloor$.
If $\bar{g} = 0$ then $\mathcal{DD}$ can be substituted by the uniform data dependence:

$$
\mathcal{DD}' = (D, U, R, \mathcal{I}_0)
$$

and the equations:

$$
\begin{aligned}
\mathbf{E}_1 &= (D, R, (V, V, \beta), f, (\mathcal{I}_0, \mathcal{I}_1, \mathcal{I}_0)) \\
\mathbf{E}_2 &= (D, \beta, in_\beta)
\end{aligned}
$$

where:

- the index mappings are:

$$
\begin{aligned}
\mathcal{I}_0(z) &= z \\
\mathcal{I}_1(z) &= z + d
\end{aligned}
$$

- R and β are new variables;
- the applied functions are:

$$
\begin{aligned}
in_\beta(z) &= g(z) \bmod 2 \\
f(a, b, c) &= \begin{cases} a & c = 0 \\ b & c = 1 \end{cases}
\end{aligned}
$$

∎ 3.2.14

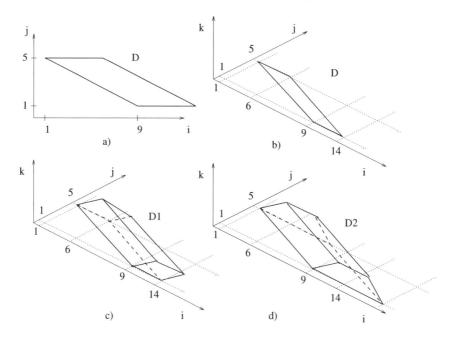

Fig. 3.10. Uniformisation: a) domain D in $\mathbf{Z}^2$ and corresponding data dependence vectors; b) domain D in $\mathbf{Z}^3$; c) domain D_1; d) domain D_2.

Example 3.2.15   Consider the atomic integral data dependence $\mathcal{DD} = (D, U, V, \mathcal{I})$ with index mapping $\mathcal{I}(i, j) = (i, j) + g(i, j)(1, 0)$, where $g(i, j) = (i^2 + j^2) mod\ 6$, and domain $D = \{(i, j) \mid 1 \leq j \leq 5, 11 - 2j \leq i \leq 16 - 2j\}$. The domain D is illustrated in Fig. 3.10 a). $lin(D) = \mathbf{Z}^2$ and $d = (1, 0) \in lin(D)$. As D is of full dimension in $\mathbf{Z}^2$, we need to reindex the data dependence in $\mathbf{Z}^3$, before applying Proposition 3.2.13 (Fig. 3.10 b) illustrates D in $\mathbf{Z}^3$). Any non-null vector in $D^\perp$ can be chosen as π. For instance, vector $\pi = (0, 0, 1)$ satisfies the conditions of Proposition 3.2.13 and $[(0, 0, 1) : 0]$ is a hyperplane containing D. Let $\hat{d} = d + \pi = (1, 0, 1)$ and $\check{d} = d - \pi = (1, 0, -1)$. For all $(i, j, k) \in D$, $g(i, j, k) \leq 5$, hence we can assume $\bar{g} = 2$. By applying Proposition 3.2.13, we obtain the data dependence

$$\mathcal{DD}' = (D, U, R^1, \mathcal{I}_0)$$

and the system of equations:

$$\mathbf{E}_1 = (D_1, R^1, (R^1, R^2, R^2, \alpha, \beta, \gamma), f, (\mathcal{I}_1, \mathcal{I}_{2,0}, \mathcal{I}_{2,1}, \mathcal{I}_0, \mathcal{I}_0, \mathcal{I}_0))$$
$$\mathbf{E}_2 = (D_{2,1}, R^2, R^2, id, \mathcal{I}_3)$$

$$
\begin{aligned}
\mathbf{E}_3 &= (D_{2,2}, R^2, V, id, \mathcal{I}_0) \\
\mathbf{E}_4 &= (D_{1,1}, \alpha, \alpha, id, \mathcal{I}_1) \\
\mathbf{E}_5 &= (D_{1,2}, \alpha, in_\alpha) \\
\mathbf{E}_6 &= (D_{1,1}, \beta, \beta, id, \mathcal{I}_1) \\
\mathbf{E}_7 &= (D_{1,2}, \beta, in_\beta) \\
\mathbf{E}_8 &= (D_{1,1}, \gamma, \gamma, dec, \mathcal{I}_1) \\
\mathbf{E}_9 &= (D_{1,2}, \gamma, in_\gamma)
\end{aligned}
$$

where:

- the index mappings are:

$$
\begin{aligned}
\mathcal{I}_0(i, j, k) &= (i, j, k) \\
\mathcal{I}_1(i, j, k) &= (i, j, k) + (1, 0, 1) \\
\mathcal{I}_{2,0}(i, j, k) &= (i, j, k) \\
\mathcal{I}_{2,1}(i, j, k) &= (i, j, k) + (1, 0, 0) \\
\mathcal{I}_3(i, j, k) &= (i, j, k) + (1, 0, -1)
\end{aligned}
$$

- the applied functions are:

$$
\begin{aligned}
in_\alpha(i, j, k) &= \lfloor g(i - 2, j, k - 2)/2 \rfloor \\
in_\beta(i, j, k) &= g(i - 2, j, k - 2) \bmod 2 \\
in_\gamma(i, j, k) &= 2 \\
id(a) &= a \\
dec(a) &= a - 1 \\
f(a, b, c, d, e, f) &= \begin{cases} a & d \neq f \\ b & d = f, e = 0 \\ c & d = f, e = 1 \end{cases}
\end{aligned}
$$

- the new domains are:

$$
\begin{aligned}
D_1 &= \{(i, j, k) \mid 1 \le j \le 5, 0 \le k \le 2, 11 - 2j + k \le i \le 16 - 2 + \\
&\quad k\} \\
D_{1,1} &= \{(i, j, k) \in D_1 \mid k < 2\} \\
D_{1,2} &= \{(i, j, k) \in D_1 \mid k = 2\} \\
D_2 &= \{(i, j, k) \mid 1 \le j \le 5, 0 \le k \le 2, 11 - 2j + k \le i \le 21 - 2j -
\end{aligned}
$$

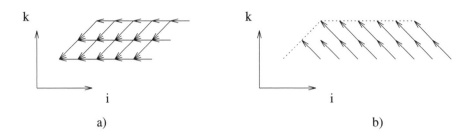

a)   b)

Fig. 3.11. Data dependence vectors after uniformisation: a) section of D_1; b) section of D_2.

$$
\begin{aligned}
& k\} \\
D_{2,1} &= \{(i,j,k) \in D_2 \mid k > 0\} \\
D_{2,2} &= \{(i,j,k) \in D_2 \mid k = 0\}.
\end{aligned}
$$

The routing domain D_1 and D_2 are sketched in Fig. 3.10 c) and d), respectively, while sections of the data dependence graph in D_1 and D_2 are given in Fig. 3.11 a) and b), respectively. As shown in the figure, the data dependence graph is uniform and its nodes are locally connected.

■ 3.2.15

3.2.4 Parametric Uniformisation

While uniformisation according to Proposition 3.2.10 defines data pipelining in the existing computation space of the problem, the technique defined by Proposition 3.2.13, in general, adds new dimensions to the specification and artificially introduces new domains for the sole purpose of routing the data. Because of the correspondence between computation points and processing elements under space-time mappings, a correspondent increase in both latency and number of processors will characterise the resulting array design. In this section we introduce a parametric version of the technique which allows us to control the amount of routing overhead to a certain extent.

The size of the routing domains defined through the uniformisation technique depends on the quantity $\bar{g} = \lfloor m/2 \rfloor$, where m is the least upper bound of the values of the coefficient g on the domain D. In this section we show that by allowing a limited amount of overloading of the data dependence graph, the size of the routing domains can be reduced. In particular, we introduce a parameter p as a small positive integer, in order to represent

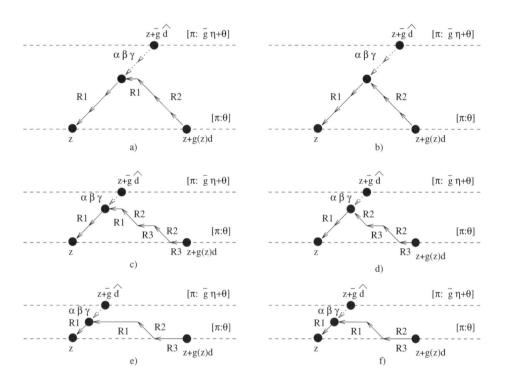

Fig. 3.12. Parametric uniformisation.

an upper bound to the amount of overloading allowed in the specification. Then we define new routing domains whose size depends on the quantity $\bar{g} = \lfloor m/(p+1) \rfloor$, and new routing paths in these smaller domains. For each z of the domain, the order of the corresponding routing path remains equal to $g(z)$. However, the shape changes according to the parameter p. Fig. 3.12 may help to illustrate how the technique works. Fig. 3.12 a) and b) illustrate the routing path (for $g(z)$ odd and even, respectively) corresponding to the data dependence vector between z and $z + g(z)d$, as defined in the previous section. For the same data dependence vector, Fig. 3.12 c) and d) and Fig. 3.12 f) and g) illustrate the corresponding routing paths for values of the parameter $p = 2$ and $p = 3$, respectively. Note that as p increases, $\bar{g}$ decreases, and the hyperplanes $[\pi : 0]$ and $[\pi : \bar{g}\eta + \theta]$ become closer together. Control variables α, β and γ are still initialised at the point $z + \bar{g}\hat{d}$. However, their initialisation values become $\lfloor g(z)/(p+1) \rfloor$, $g(z) \bmod (p+1)$ and $\bar{g}$, respectively. As before, the control variables are pipelined according to the direction of $-\hat{d}$. At each step the value of γ is decreased by one, and α and γ are compared. When their values are equal, the value of β determines at which lattice point the data has to be transferred between R^2 and R^1. Note that three variables R^1, R^2 and R^3 are used to route the data. The figure illustrates the correspondence between the routing variables and the routing directions d, $\hat{d}$ and $\check{d}$.

Proposition 3.2.16 *[Parametric Uniformisation]* Let $\mathcal{DD} = (D, U, V, \mathcal{I})$ be an atomic integral data dependence, with $\mathcal{I}(z) = z + g(z)d$. Let g be non-negative and bounded over D, and m the least upper bound of g. Let $p \in \mathbf{N}^+$ and $\bar{g} = \lfloor m/(p+1) \rfloor$. Let $d \in lin(D)$ and $dim(D) < n$. Consider $\pi \in D^\perp$, with $\pi \neq 0$, the hyperplane $[\pi : 0]$ containing the domain D, and let $\hat{d} = d + \pi$ and $\check{d} = d - \pi$.

Then $\mathcal{DD}$ can be substituted by the uniform data dependence

$$\mathcal{DD}' = (D, U, R^1, \mathcal{I}_0)$$

and the system of equations:

$$
\begin{aligned}
\mathbf{E}_1 &= (D_1, R^1, (R^1, R^2, \ldots, R^2, \alpha, \beta, \gamma), f, (\mathcal{I}_1, \mathcal{I}_{2,0}, \ldots, \mathcal{I}_{2,p}, \mathcal{I}_0, \mathcal{I}_0, \mathcal{I}_0)) \\
\mathbf{E}_2 &= (D_{2,1}, R^2, R^3, id, \mathcal{I}_3) \\
\mathbf{E}_3 &= (D_{2,2}, R^2, V, id, \mathcal{I}_0) \\
\mathbf{E}_4 &= (D_{2,3}, R^3, R^2, id, \mathcal{I}_4) \\
\mathbf{E}_5 &= (D_{1,1}, \alpha, \alpha, id, \mathcal{I}_1)
\end{aligned}
$$

$$\mathbf{E}_6 = (D_{1,2}, \alpha, in_\alpha)$$
$$\mathbf{E}_7 = (D_{1,1}, \beta, \beta, id, \mathcal{I}_1)$$
$$\mathbf{E}_8 = (D_{1,2}, \beta, in_\beta)$$
$$\mathbf{E}_9 = (D_{1,1}, \gamma, \gamma, dec, \mathcal{I}_1)$$
$$\mathbf{E}_{10} = (D_{1,2}, \gamma, in_\gamma)$$

where:

- the index mappings are:

$$\mathcal{I}_0(z) = z$$
$$\mathcal{I}_1(z) = z + \hat{d}$$
$$\mathcal{I}_{2,0}(z) = z$$
$$\mathcal{I}_{2,1}(z) = z + d$$
$$\cdots$$
$$\mathcal{I}_{2,p}(z) = z + pd$$
$$\mathcal{I}_3(z) = z + \check{d}$$
$$\mathcal{I}_4(z) = z + (p - 1)d$$

- $R^1, R^2, R^3, \alpha, \beta$ and γ are new variables;

- the applied functions are:

$$in_\alpha(z) = \lfloor g(z - \bar{g}\hat{d})/(p + 1) \rfloor$$
$$in_\beta(z) = g(z - \bar{g}\hat{d}) \bmod (p + 1)$$
$$in_\gamma(z) = \bar{g}$$
$$id(a) = a$$
$$dec(a) = a - 1$$
$$f(a, b_0, \ldots, b_p, c, d, e) = \begin{cases} a & c \neq e \\ b_0 & c = e, d = 0 \\ \cdots \\ b_p & c = e, d = p \end{cases}$$

- the new domains are:

$$D_1 = \{z + l\hat{d} \mid z \in D, 0 \leq l \leq \bar{g}\}$$
$$D_{1,1} = \{z + l\hat{d} \mid z \in D, 0 \leq l < \bar{g}\}$$

$$\begin{aligned}
D_{1,2} &= \{z + \bar{g}\hat{d} \mid z \in D\} \\
D_2 &= \{z + (l_1 + l_2)d + l_3\check{d} \mid z \in D_1, 0 \le l_1 \le p, 0 \le l_2 \le (p-1)\bar{g}, \\
&\quad 0 \le l_3 \le \bar{g}\} \cap \{z \in \mathbf{Z}^n \mid \pi \cdot z \ge \theta\} \\
D_{2,1} &= \{z \in D_2 \mid \pi \cdot z > \theta\} \\
D_{2,2} &= \{z \in D_2 \mid \pi \cdot z = \theta\} \\
D_{2,3} &= \{z \in D_2 \mid \pi \cdot z < \theta + \bar{g}\eta\}.
\end{aligned}$$

PROOF: For all $z \in D$, let $segm(z) = \{z + ld \mid 0 \le l \le \bar{g}\}$. By definition, for all $z' \in segm(z)$, $\alpha(z') = \lfloor g(z)/(p+1)\rfloor$. In fact, for $z \in D$,

$$\begin{aligned}
\alpha(z) = \alpha(z + \hat{d}) &= \ldots = \alpha(z + \bar{g}\hat{d}) \\
&= \lfloor g(z + \bar{g}\hat{d} - \bar{g}\hat{d})/(p+1)\rfloor = \lfloor g(z)/(p+1)\rfloor.
\end{aligned}$$

Also, for all $z' = z + l\hat{d}$ in the segment $segm(z)$, with $z \in D$, $\beta(z') = g(z) \bmod (p+1)$. In fact, for $z \in D$,

$$\begin{aligned}
\beta(z) = \beta(z + \hat{d}) &= \ldots = \beta(z + \bar{g}\hat{d}) \\
&= g(z + \bar{g}\hat{d} - \bar{g}\hat{d}) \bmod (p+1) = g(z) \bmod (p+1).
\end{aligned}$$

Finally, for all $z' = z + l\hat{d} \in segm(z)$, $\gamma(z') = l$. In fact, for $z \in D$,

$$\begin{aligned}
\gamma(z) = \gamma(z + \hat{d}) &- 1 \\
&= \ldots = \gamma(z + \bar{g}\hat{d}) - \bar{g} = \bar{g} - \bar{g} = 0.
\end{aligned}$$

Therefore, for all $z' = z + ld \in segm(z)$, $\alpha(z') = \gamma(z')$ if and only if $l = \lfloor g(z)/(p+1)\rfloor$.

We observe that, for all $c \in \mathbf{Z}$,

$$\begin{aligned}
\lfloor c/(p+1)\rfloor &= (c - c \bmod (p+1))/(p+1) \\
c &= (p+1)\lfloor c/(p+1)\rfloor + c \bmod (p+1).
\end{aligned}$$

Hence, for all $z \in D$,

$$\begin{aligned}
U(z) = R^1(\mathcal{I}_0(z)) &= R^1(\mathcal{I}_1 \circ \mathcal{I}_0(z)) \\
&= \ldots = R^1(\mathcal{I}_1^{\lfloor g(z)/(p+1)\rfloor} \circ \mathcal{I}_0(z)) \\
&= R^2(\mathcal{I}_{2,(g(z) \bmod (p+1))} \circ \mathcal{I}_1^{\lfloor g(z)/(p+1)\rfloor} \circ \mathcal{I}_0(z)) \\
&= R^3(\mathcal{I}_3 \circ \mathcal{I}_{2,(g(z) \bmod (p+1))} \circ \mathcal{I}_1^{\lfloor g(z)/(p+1)\rfloor} \circ \mathcal{I}_0(z))
\end{aligned}$$

$$= R^2(\mathcal{I}_4 \circ \mathcal{I}_3 \circ \mathcal{I}_{2,(g(z) \bmod (p+1))} \circ \mathcal{I}_1^{\lfloor g(z)/(p+1)\rfloor} \circ \mathcal{I}_0(z))$$

$$= \ldots$$

$$= R^2(\mathcal{I}_4^{\lfloor g(z)/(p+1)\rfloor} \circ \mathcal{I}_3^{\lfloor g(z)/(p+1)\rfloor} \circ \mathcal{I}_{2,(g(z) \bmod (p+1))}$$
$$\circ \mathcal{I}_1^{\lfloor g(z)/(p+1)\rfloor} \circ \mathcal{I}_0(z))$$

$$= V(\mathcal{I}_0 \circ \mathcal{I}_4^{\lfloor g(z)/(p+1)\rfloor} \circ \mathcal{I}_3^{\lfloor g(z)/(p+1)\rfloor} \circ \mathcal{I}_{2,(g(z) \bmod (p+1))}$$
$$\circ \mathcal{I}_1^{\lfloor g(z)/(p+1)\rfloor} \circ \mathcal{I}_0(z))$$

$$= V(z + \lfloor g(z)/(p+1)\rfloor\hat{d} + (g(z) \bmod (p+1))d +$$
$$\lfloor g(z)/(p+1)\rfloor\check{d} + \lfloor g(z)/(p+1)\rfloor(p-1)d)$$

$$= V(z + ((p+1)\lfloor g(z)/(p+1)\rfloor + g(z) \bmod (p+1))d)$$

$$= V(z + g(z)d) = V(\mathcal{I}(z)).$$

∎ 3.2.16

Corollary 3.2.17 Let $\mathcal{DD} = (D, U, V, \mathcal{I})$ be an integral data dependence, with $\mathcal{I}(z) = z + g(z)d$. Let g be non-negative and bounded over D, and m the least upper bound of g. Let $p \in \mathbf{N}^+$ and $\bar{g} = \lfloor m/(p+1)\rfloor$.

If $\bar{g} = 0$ then $\mathcal{DD}$ can be substituted by the uniform data dependence:

$$\mathcal{DD}' = (D, U, R, \mathcal{I}_0)$$

and the equations:

$$\mathbf{E}_1 = (D, R, (V, \ldots, V, \beta), f, (\mathcal{I}_{1,0}, \ldots, \mathcal{I}_{1,p}, \mathcal{I}_0))$$
$$\mathbf{E}_2 = (D, \beta, in_\beta)$$

where:

- the index mappings are:

$$\mathcal{I}_0(z) = z$$
$$\mathcal{I}_{1,0}(z) = z$$
$$\mathcal{I}_{1,1}(z) = z + d$$
$$\ldots$$
$$\mathcal{I}_{1,p}(z) = z + pd$$

- R and β are new variables;

- the applied functions are:

$$in_\beta(z) \quad = \quad g(z) \; mod \; (p+1)$$

$$f(a_0, \ldots, a_p, b) \quad = \quad \begin{cases} a_0 & b = 0 \\ a_1 & b = 1 \\ \cdots & \\ a_p & b = p \end{cases}$$

$\blacksquare$ 3.2.17

Note that uniformisation according to Proposition 3.2.13 corresponds to parametric uniformisation with parameter $p = 1$.

Example 3.2.18   Let us apply parametric uniformisation to the atomic integral data dependence of Example 3.2.15, with the same choice of routing directions, i.e., $d = (1, 0, 0)$, $\hat{d} = (1, 0, 1)$ and $\check{d} = (1, 0, -1)$.

Let $\bar{g} = \lfloor 5/(p+1) \rfloor$, with $p \in \mathbf{N}^+$. The application of parametric uniformisation according to Proposition 3.2.16 yields the data dependence

$$\mathcal{DD}' \quad = \quad (D, U, R^1, \mathcal{I}_0)$$

and the system of equations:

$$
\begin{aligned}
\mathbf{E}_1 &= (D_1, R^1, (R^1, R^2, \ldots, R^2, \alpha, \beta, \gamma), f, (\mathcal{I}_1, \mathcal{I}_{2,0}, \ldots, \mathcal{I}_{2,p}, \mathcal{I}_0, \mathcal{I}_0, \mathcal{I}_0)) \\
\mathbf{E}_2 &= (D_{2,1}, R^2, R^3, id, \mathcal{I}_3) \\
\mathbf{E}_3 &= (D_{2,2}, R^2, V, id, \mathcal{I}_0) \\
\mathbf{E}_4 &= (D_{2,3}, R^3, R^2, id, \mathcal{I}_4) \\
\mathbf{E}_5 &= (D_{1,1}, \alpha, \alpha, id, \mathcal{I}_1) \\
\mathbf{E}_6 &= (D_{1,2}, \alpha, in_\alpha) \\
\mathbf{E}_7 &= (D_{1,1}, \beta, \beta, id, \mathcal{I}_1) \\
\mathbf{E}_8 &= (D_{1,2}, \beta, in_\beta) \\
\mathbf{E}_9 &= (D_{1,1}, \gamma, \gamma, dec, \mathcal{I}_1) \\
\mathbf{E}_{10} &= (D_{1,2}, \gamma, in_\gamma)
\end{aligned}
$$

where:

- the index mappings are:

$$
\begin{aligned}
\mathcal{I}_0(i, j, k) &= (i, j, k) \\
\mathcal{I}_1(i, j, k) &= (i, j, k) + (1, 0, 1)
\end{aligned}
$$

$$
\begin{aligned}
\mathcal{I}_{2,0}(i,j,k) &= (i,j,k) \\
\mathcal{I}_{2,1}(i,j,k) &= (i,j,k) + (1,0,0) \\
&\cdots \\
\mathcal{I}_{2,p}(i,j,k) &= (i,j,k) + (p,0,0) \\
\mathcal{I}_{3}(i,j,k) &= (i,j,k) + (1,0,-1) \\
\mathcal{I}_{4}(i,j,k) &= (i,j,k) + (p-1,0,0)
\end{aligned}
$$

- the applied functions are:

$$
\begin{aligned}
in_{\alpha}(i,j,k) &= \lfloor g(i-\bar{g},j,k-\bar{g})/(p+1) \rfloor \\
in_{\beta}(i,j,k) &= g(i-\bar{g},j,k-\bar{g}) \bmod (p+1) \\
in_{\gamma}(i,j,k) &= \bar{g} \\
id(a) &= a \\
dec(a) &= a-1 \\
f(a,b_0,\ldots,b_p,c,d,e) &= \begin{cases} a & c \neq e \\ b_0 & c = e, d = 0 \\ \cdots \\ b_p & c = e, d = p \end{cases}
\end{aligned}
$$

- the new domains are:

$$
\begin{aligned}
D_1 &= \{(i,j,k) \mid 1 \le j \le 5, 0 \le k \le \bar{g}, 11 - 2j + k \le i \le 16 - 2j + k\} \\
D_{1,1} &= \{(i,j,k) \in D_1 \mid k < \bar{g}\} \\
D_{1,2} &= \{(i,j,k) \in D_1 \mid k = \bar{g}\} \\
D_2 &= \{(i,j,k) \mid 1 \le j \le 5, 0 \le k \le \bar{g}, 11 - 2j + k \le i \le 21 - 2j - k\} \\
D_{2,1} &= \{(i,j,k) \in D_2 \mid k > 0\} \\
D_{2,2} &= \{(i,j,k) \in D_2 \mid k = 0\} \\
D_{2,3} &= \{(i,j,k) \in D_2 \mid k < \bar{g}\}.
\end{aligned}
$$

The routing domains D_1 and D_2 are similar to those sketched in Fig. 3.10 (as the same regularisation vectors are considered). A section of the data dependence graph in D_1 and D_2 is given in Fig. 3.13 a), b) and Fig. 3.13 c), d) for values of the parameter $p = 2$ and $p = 3$, respectively.   ■ 3.2.18

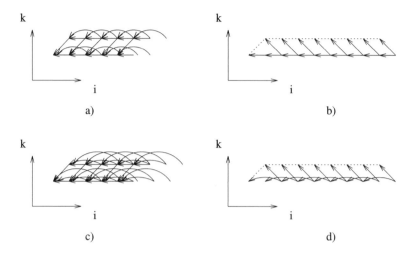

Fig. 3.13. Data dependence vectors after parametric uniformisation: a) and b) sections of D_1 and D_2 for $p = 2$; c) and d) the same sections for $p = 3$.

3.2.5   Decomposition

While uniformisation replaces an atomic integral data dependence with a system of uniform recurrences, decomposition allows us to substitute an integral data dependence with a set of atomic integral data dependencies. This is obtained by decomposing all data dependence vectors into their components along the generators of the integral dependence mapping.

Given an integral data dependence $\mathcal{DD} = (D, U, V, \mathcal{I})$, whose integral index mapping $\mathcal{I}$ has m generators, where $m > 1$, the application of a decomposition technique aims at generating a set of integral data dependencies such that:

- each of the new index mappings is integral with less than m generators; and

- the composition of the new index mappings is equal to $\mathcal{I}$ for each point of the domain D.

A recursive application of the techniques yields a system of atomic integral data dependencies. Decomposition is subject to the same constraint as uniformisation, namely that the coefficients of the index mapping define bounded integer functions on the computation domain. Also, similar to uniformisation, two decomposition techniques can be defined according to the

Fig. 3.14. Decomposition 1.

relation between the generators of the index mapping and the domain of the data dependence.

In the first case, illustrated in Fig. 3.14 (in 2 dimensions), the application of the technique consists of:

- the selection of a generator d of $\mathcal{I}$ such that $d \notin lin(D)$. Let g be the coefficient of $\mathcal{I}$ relative to d;

- the definition of the atomic integral index mapping $\mathcal{I}_0(z) = z + g(z)d$. Because of Proposition 3.2.4, $\mathcal{I}_0$ is guaranteed to be injective over D and an inverse $\mathcal{I}_0^{-1}$ can be defined according to Proposition 3.2.7;

- the definition of a new integral index mapping $\mathcal{I}_1$, based on the remaining generators of $\mathcal{I}$, and having as coefficients the composition of $\mathcal{I}_0^{-1}$ with the relative coefficients of $\mathcal{I}$.

Suitable routing domains and an auxiliary variable are introduced. Fig. 3.14 a) and b) illustrate the decomposition technique for $\mathcal{I}(z)$, respectively, inside and outside $lin(D)$.

Proposition 3.2.19 [*Decomposition 1*] Let $\mathcal{DD} = (D, U, V, \mathcal{I})$ be an integral data dependence, with $\mathcal{I}(z) = z + g(z)d + \sum_{j=1}^{m} g_j(z)d_j$. Let g be non-negative and bounded over D, and $\bar{g}$ the least upper bound of g. Let $d \notin lin(D)$ and P be the convex polyhedron generated by D and d. Consider $\pi \in lin(P) \cap D^{\perp}$, with $\pi \neq 0$, and the hyperplane $[\pi : 0]$ containing the domain D.

Then $\mathcal{DD}$ can be substituted by the atomic integral data dependence

$$\mathcal{DD}' = (D, U, R, \mathcal{I}_0)$$

and the equation

$$\mathbf{E} = (D_1, R, V, id, \mathcal{I}_1)$$

where:

- the index mappings are:

$$\mathcal{I}_0(z) \;=\; z + g(z)d$$

$$\mathcal{I}_1(z) \;=\; z + \sum_{j=1}^{m} g'_j(z)d_j$$

with $g'_j(z) = g_j(\mathcal{I}_0^{-1}(z))$, $\mathcal{I}_0^{-1}(z) = z + l(z)(-d)$, $l(z) = (\pi \cdot z - \theta)/\eta$ and $\eta = \pi \cdot d$;

- R is a new variable;

- the applied function is $id(a) = a$;

- the new domain is $D_1 = \{z + ld \mid z \in D, 0 \le l \le \bar{g}\}$.

PROOF: $\mathcal{I}_0^{-1}$ defines an inverse of $\mathcal{I}_0$ on D, according to Proposition 3.2.7. Hence, for all $z \in D$, $\mathcal{I}_0^{-1} \circ \mathcal{I}_0(z) = z$.

Therefore, for all $z \in D$,

$$\mathcal{I}_1 \circ \mathcal{I}_0(z) =$$
$$= \mathcal{I}_0(z) + \sum_{j=1}^{m} g'_j(\mathcal{I}_0(z))d_j$$
$$= \mathcal{I}_0(z) + \sum_{j=1}^{m} g_j(\mathcal{I}_0^{-1} \circ \mathcal{I}_0(z))d_j$$
$$= z + g(z)d + \sum_{j=1}^{m} g_j(z)d_j = \mathcal{I}(z)$$

Finally, for all $z \in D$,

$$U(z) = R(\mathcal{I}_0(z))$$
$$= V(\mathcal{I}_1 \circ \mathcal{I}_0(z)) = V(\mathcal{I}(z)).$$

■ 3.2.19

Example 3.2.20   Consider the integral data dependence $\mathcal{DD} = (D, U, V, \mathcal{I})$ with index mapping $\mathcal{I}(i, j) = (i, j) + g_1(i, j)(1, 0) + g_2(i, j)(0, 1)$, where $g_1(i, j) = (i^2 + j^2) \bmod 6$ and $g_2(i, j) = 2^j$, and domain $D = \{(i, j) \mid 1 \leq j \leq 5, i = 11 - 2j\}$. $lin(D) = \langle(2, -1)\rangle$ and $d = (1, 0) \notin lin(D)$.

We can choose π as any vector in $D^\perp$. In particular $\pi = (1, 2)$ (which is orthogonal to the generator $(2, -1)$ of $lin(D)$) satisfies the conditions of Proposition 3.2.19. Then, $[(1, 2) : 11]$ is a hyperplane containing D, $\eta = (1, 2) \cdot (1, 0) = 1$ and $l(i, j) = i + 2j - 11$. Also, for all $(i, j) \in D$, $g_1(i, j) \leq 5$. If we apply Proposition 3.2.19, we obtain the data dependence

$$\mathcal{DD}' \;=\; (D, U, R, \mathcal{I}_0)$$

and the equation

$$\mathbf{E} \;=\; (D_1, R, V, id, \mathcal{I}_1)$$

where $id(a) = a$, the index mappings are:

$$\begin{aligned}
\mathcal{I}_0(i, j) &= (i, j) + g_1(i, j)(1, 0) \\
\mathcal{I}_1(i, j) &= (i, j) + g_2'(i, j)(0, 1)
\end{aligned}$$

and the new domain is $D_1 = \{(i, j) \mid 1 \leq j \leq 5, 11 - 2j \leq i \leq 16 - 2j\}$. Note the new coefficient g_2', which is defined, according to the proposition, as the composition of g_2 and $\mathcal{I}_0^{-1}$, that is:

$$\begin{aligned}
\mathcal{I}_0^{-1}(i, j) &= (i, j) + l(i, j)(-1, 0) \\
g_2'(i, j) &= g_2(\mathcal{I}_0^{-1}(i, j)) = 2^j
\end{aligned}$$

∎ 3.2.20

The second decomposition technique is similar to the first one, and corresponds to a situation in which the selected generator d of $\mathcal{I}$ is contained in the direction of the domain D, i.e., $d \in lin(D)$. As for uniformisation, a vector π needs to be chosen so that routing directions outside $lin(D)$ can be considered. The choice of π is that of a vector orthogonal to both $lin(D)$ and the space spanned by the generators of $\mathcal{I}$. The resulting decomposition is illustrated (in 3 dimensions) in Fig. 3.15 a) and b) for $\mathcal{I}(z)$ inside and outside $lin(D)$, respectively. Note that this choice of π implies that even for a 1-dimensional domain D, the application of the technique may require a 3-dimensional space (see Fig. 3.15 b)). The direction vectors $\hat{d} = d + \pi$ and $-\pi$ are considered as the generators of the new integral index mappings. Let g be the coefficient of $\mathcal{I}$ relative to d. The technique consists of:

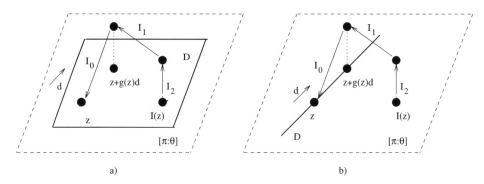

Fig. 3.15. Decomposition 2.

- the definition of the atomic integral index mapping $\mathcal{I}_0 = z + g(z)\hat{d}$. Because of Proposition 3.2.4 and the choice of π, $\hat{d}$ is not in $lin(D)$, the mapping $\mathcal{I}_0$ is guaranteed to be injective over D, and an inverse $\mathcal{I}_0^{-1}$ can be defined according to Proposition 3.2.7;

- the definition of an integral index mapping $\mathcal{I}_1$ based on the remaining generators of $\mathcal{I}$, and having as coefficients the composition of $\mathcal{I}_0^{-1}$ with the relative coefficients of $\mathcal{I}$; and

- the definition of a third atomic integral index mapping $\mathcal{I}_2$ with generator $-\pi$.

Proposition 3.2.21 *[Decomposition 2]* Let $\mathcal{DD} = (D, U, V, \mathcal{I})$ be an integral data dependence, with $\mathcal{I}(z) = z + g(z)d + \sum_{j=1}^{m} g_j(z)d_j$. Let g, g_j be non-negative and bounded over D and let $\bar{g}, \bar{g}_j$ be the least upper bounds of g, g_j, respectively. Let $d \in lin(D)$, P be the convex polyhedron generated by D and $\{d_1, \ldots, d_m\}$, and $dim(P) < n$. Consider $\pi \in P^{\perp}$, with $\pi \neq 0$, the hyperplane $[\pi : \theta]$ containing the domain D, and let $\hat{d} = d + \pi$.

Then $\mathcal{DD}$ can be substituted by the atomic integral data dependence

$$\mathcal{DD}' = (D, U, R^1, \mathcal{I}_0)$$

and the equations:

$$\mathbf{E}_1 = (D_1, R^1, R^2, id, \mathcal{I}_1)$$
$$\mathbf{E}_2 = (D_2, R^2, V, id, \mathcal{I}_2)$$

where:

- the index mappings are:

$$\mathcal{I}_0(z) \;=\; z + g(z)\hat{d}$$

$$\mathcal{I}_1(z) \;=\; z + \sum_{j=1}^{m} g_j'(z)d_j$$

$$\mathcal{I}_2(z) \;=\; z + g'(z)(-\pi)$$

with $g_j'(z) = g_j(\mathcal{I}_0^{-1}(z))$, $\mathcal{I}_0^{-1}(z) = z + l(z)(-\hat{d})$, $g'(z) = l(z)$, $l(z) = (\pi \cdot z - \theta)/\eta$ and $\eta = \pi \cdot \hat{d}$;

- R^1, R^2 are new variables;

- the applied function is $id(a) = a$;

- the new domains are:

$$D_1 \;=\; \{z + l\hat{d} \mid z \in D, 0 \leq l \leq \bar{g}\}$$

$$D_2 \;=\; \{z + \sum_{j=1}^{m} l_j d_j \mid z \in D_1, 0 \leq l_j \leq \bar{g}_j\}.$$

PROOF: As $\pi \notin lin(D)$, then $\hat{d} = d + \pi \notin lin(D)$ and $\mathcal{I}_0$ is injective over D. Hence, $\mathcal{I}_0^{-1}$ defines an inverse of $\mathcal{I}_0$ on D, according to Proposition 3.2.7. Therefore, for all $z \in D$, $\mathcal{I}_0^{-1} \circ \mathcal{I}_0(z) = z$.

For all $z \in D$,

$$\mathcal{I}_1 \circ \mathcal{I}_0(z) =$$

$$= \mathcal{I}_0(z) + \sum_{j=1}^{m} g_j'(\mathcal{I}_0(z))d_j$$

$$= \mathcal{I}_0(z) + \sum_{j=1}^{m} g_j(\mathcal{I}_0^{-1} \circ \mathcal{I}_0(z))d_j$$

$$= z + g(z)\hat{d} + \sum_{j=1}^{m} g_j(z)d_j.$$

Also, as $\pi \in P^{\perp}$, then $\pi \cdot d_j = 0$ and $\pi \cdot (\sum_{j=1}^{m} g_j(z)d_j) = 0$. Hence, for all $z \in D$,

$$g'(\mathcal{I}_1 \circ \mathcal{I}_0(z)) = l(\mathcal{I}_1 \circ \mathcal{I}_0(z))$$

$$\begin{aligned}
&= \ (\pi \cdot (\mathcal{I}_1 \circ \mathcal{I}_0(z)) - \theta)/\eta \\
&= \ (\pi \cdot (z + g(z)\hat{d} + \sum_{j=1}^{m} g_j(z)d_j) - \theta)/\eta \\
&= \ (\pi \cdot z + g(z)\pi \cdot \hat{d} + \pi \cdot (\sum_{j=1}^{m} g_j(z)d_j) - \theta)/\eta \\
&= \ (\theta + g(z)\eta + 0 - \theta)/\eta = g(z).
\end{aligned}$$

Therefore, for all $z \in D$,

$$\begin{aligned}
\mathcal{I}_2 \circ \mathcal{I}_1 \circ \mathcal{I}_0(z) &= \mathcal{I}_2(\mathcal{I}_1 \circ \mathcal{I}_0(z)) \\
&= \ \mathcal{I}_1 \circ \mathcal{I}_0(z) + g'(\mathcal{I}_1 \circ \mathcal{I}_0(z))(-\pi) \\
&= \ \mathcal{I}_1 \circ \mathcal{I}_0(z) + g(z)(-\pi) \\
&= \ z + g(z)\hat{d} + \sum_{j=1}^{m} g_j(z)d_j + g(z)(-\pi) \\
&= \ z + g(z)(\pi + d - \pi) + \sum_{j=1}^{m} g_j(z)d_j \\
&= \ z + g(z)d + \sum_{j=1}^{m} g_j(z)d_j = \mathcal{I}(z).
\end{aligned}$$

Finally, for all $z \in D$,

$$\begin{aligned}
U(z) = R^1(\mathcal{I}_0(z)) &= R^2(\mathcal{I}_1 \circ \mathcal{I}_0(z)) \\
&= \ V(\mathcal{I}_2 \circ \mathcal{I}_1 \circ \mathcal{I}_0(z)) = V(\mathcal{I}(z)).
\end{aligned}$$

$\blacksquare$ 3.2.21

Condition $dim(P) < n$ above implies that it may be necessary to increase the number of dimensions of the space prior to the application of Proposition 3.2.21.

Example 3.2.22   Consider the integral data dependence $\mathcal{DD} = (D, U, V, \mathcal{I})$ with index mapping $\mathcal{I}(i, j) = (i, j) + g_1(i, j)(1, 0) + g_2(i, j)(0, 1)$, where $g_1(i, j) = (i^2 + j^2)mod\ 6$ and $g_2(i, j) = 2^j$, and domain $D = \{(i, j) \mid 1 \le j \le 5, 11 - 2j \le i \le 16 - 2j\}$. $lin(D) = \mathbf{Z}^2$ and $d = (1, 0) \in lin(D)$. As the polyhedron P generated by D and $d_2 = (0, 1)$ is of full dimension in $\mathbf{Z}^2$, we need to reindex the data dependence in $\mathbf{Z}^3$, before applying Proposition 3.2.21.

Any non-null vector in $D^\perp$ can be chosen as π. For instance, vector $\pi = (0,0,1)$ satisfies the conditions of Proposition 3.2.21 and $[(0,0,1):0]$ is a hyperplane containing P. Let $\hat{d} = d + \pi = (1,0,1)$, $\eta = (0,0,1) \cdot (1,0,1) = 1$ and $l(i,j,k) = k$. Note that, for all $(i,j,k) \in D$, $g_1(i,j,k) \leq 5$, and $g_2(i,j,k) \leq 2^5$. If we apply Proposition 3.2.21 we obtain the data dependence

$$\mathcal{DD}' = (D, U, R^1, \mathcal{I}_0)$$

and the equations:

$$\mathbf{E}_1 = (D_1, R^1, R^2, id, \mathcal{I}_1)$$
$$\mathbf{E}_2 = (D_2, R^2, V, id, \mathcal{I}_2)$$

where:

- the index mappings are:

$$\mathcal{I}_0(i,j,k) = (i,j,k) + g_1(i,j,k)(1,0,1)$$
$$\mathcal{I}_1(i,j,k) = (i,j,k) + g_2'(i,j,k)(0,1,0)$$
$$\mathcal{I}_2(i,j,k) = (i,j,k) + g'(i,j,k)(0,0,-1)$$

- the new domains are:

$$D_1 = \{(i,j,k) \mid 1 \leq j \leq 5, 0 \leq k \leq 5, 11 - 2j + k \leq i \leq 16 - 2j + k\}$$
$$D_2 = \{(i,j,k) \mid 16 \leq i + 2j - k \leq 16 + 2^6, 1 \leq 1i - k \leq 14, 1 \leq j \leq 5 + 2^5\}.$$

Note the new coefficients g_2' and g' defined, according to Proposition 3.2.21, as:

$$g_2'(i,j,k) = g_2(\mathcal{I}_0^{-1}(i,j,k)) = 2^j$$
$$g'(i,j,k) = l(i,j,k) = k$$

where $\mathcal{I}_0^{-1}(i,j,k) = (i,j,k) + l(i,j,k)(-1,0,-1) = (i-k,j,0)$. ■ 3.2.22

The previous results provide means of defining decomposition techniques for integral data dependencies which can be automated. The decomposition procedure is based on the possibility of defining, at each step, an inverse of an atomic integral index mapping. This definition relies upon the results in Propositions 3.2.4 and 3.2.7, and, in general, may require an increase in the dimensionality of the computation space. A more economical decomposition step may be defined if such an inverse is known by the algorithm designer. In such a case, a decomposition technique may be defined as follows:

Proposition 3.2.23 *[Decomposition 3]* Let $\mathcal{DD} = (D, U, V, \mathcal{I})$ be an integral data dependence, with $\mathcal{I}(z) = z + g(z)d + \sum_{j=1}^{m} g_j(z)d_j$. Let g be non-negative and bounded over D, and $\bar{g}$ the least upper bound of g. Let $\mathcal{I}_0(z) = z + g(z)d$ be an injective mapping over D with inverse $\mathcal{I}_0^{-1}$.

Then $\mathcal{DD}$ can be substituted by the atomic integral data dependence

$$\mathcal{DD}' \;=\; (D, U, R, \mathcal{I}_0)$$

and the equation

$$\mathbf{E} \;=\; (D_1, R, V, id, \mathcal{I}_1)$$

where:

- the index mappings are:

$$\mathcal{I}_0(z) \;=\; z + g(z)d$$

$$\mathcal{I}_1(z) \;=\; z + \sum_{j=1}^{m} g'_j(z)d_j$$

   with $g'_j(z) = g_j(\mathcal{I}_0^{-1}(z))$;

- R is a new variable;

- the applied function is $id(a) = a$;

- the new domain is $D_1 = \{z + ld \mid z \in D, 0 \leq l \leq \bar{g}\}$.

PROOF: As $\mathcal{I}_0^{-1}$ is an inverse of $\mathcal{I}_0$ on D, then for all $z \in D$, $\mathcal{I}_0^{-1} \circ \mathcal{I}_0(z) = z$. Therefore, for all $z \in D$,

$$\mathcal{I}_1 \circ \mathcal{I}_0(z) =$$

$$= \; \mathcal{I}_0(z) + \sum_{j=1}^{m} g'_j(\mathcal{I}_0(z))d_j$$

$$= \; \mathcal{I}_0(z) + \sum_{j=1}^{m} g_j(\mathcal{I}_0^{-1} \circ \mathcal{I}_0(z))d_j$$

$$= \; z + g(z)d + \sum_{j=1}^{m} g_j(z)d_j = \mathcal{I}(z).$$

Finally, for all $z \in D$,

$$\begin{aligned} U(z) &= R(\mathcal{I}_0(z)) \\ &= V(\mathcal{I}_1 \circ \mathcal{I}_0(z)) = V(\mathcal{I}(z)). \end{aligned}$$

∎ 3.2.23

Example 3.2.24 Consider the integral data dependence $\mathcal{DD} = (D, U, V, \mathcal{I})$ with index mapping $\mathcal{I}(i,j) = (i,j) + (1,1) + g(i,j)(1,0)$, where $g(i,j) = (i^2 + j^2) \bmod 6$, and domain $D = \{(i,j) \mid 1 \le j \le 5, 11 - 2j \le i \le 16 - 2j\}$.
   The index mapping $\mathcal{I}_0(i,j) = (i,j) + (1,1)$ admits the inverse $\mathcal{I}_0^{-1}(i,j) = (i,j) + (-1,-1)$. Hence, $\mathcal{DD}$ can be substituted by the data dependence

$$\mathcal{DD}' = (D, U, R, \mathcal{I}_0)$$

and the equation

$$\mathbf{E} = (D_1, R, V, id, \mathcal{I}_1)$$

where $\mathcal{I}_1(i,j) = (i,j) + g'(i,j)(1,0)$, $g'(i,j) = g(\mathcal{I}_0^{-1}(i,j)) = ((i-1)^2 + (j - 1)^2) \bmod 6$, and $D_1 = \{(i+1, j+1) \mid (i,j) \in D\}$.
   Note that as $lin(D) = \mathbf{Z}^2$ and $d = (1,0) \in lin(D)$, the application of Proposition 3.2.21 would require an increase in the dimensionality of the space. ∎ 3.2.24

From the previous results, we obtain the corollary:

Corollary 3.2.25 Let $\mathcal{DD} = (D, U, V, \mathcal{I})$ be an integral data dependence, with $\mathcal{I}(z) = z + g(z)d + \sum_{j=1}^{m} g_j(z)d_j$. Let g be non-negative and bounded over D, and $\bar{g}$ the least upper bound of g.
   If $\bar{g} = 0$, then $\mathcal{DD}$ can be substituted by the integral data dependence

$$\mathcal{DD}_0 = (D, U, V, \mathcal{I}_0)$$

where $\mathcal{I}_0(z) = z + \sum_{j=1}^{m} g_j(z)d_j$. ∎ 3.2.25

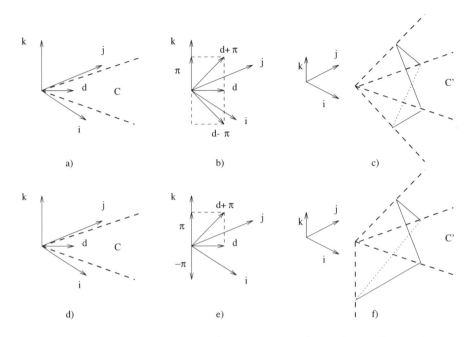

Fig. 3.16. The preservation of pointedness through regularisation.

3.3   Regularisation and Affine Scheduling

Form the discussion in Section 2.2.6, a pointed dependence cone implies the existence of valid affine timing functions. The preservation of such functions by regularisation can be guaranteed if the techniques realise transformations of pointed dependence cones.

The following two propositions provide the basic results to prove that integral regularisation techniques preserve affine scheduling. The results are illustrated in Fig. 3.16 for a 3-dimensional space. Parts a), b) and c) of the figure illustrate Proposition 3.3.1, while parts d), e) and f) correspond to Proposition 3.3.2.

Proposition 3.3.1   Let C be a pointed polyhedral convex cone with generators $r_1, \ldots, r_p$. Let $\rho \in C^\perp$ and $r \in C$, with $\rho \neq 0$ and $r \neq 0$. Define $\hat{r} = r + \rho$ and $\check{r} = r - \rho$. Then the cone C' generated by $r_1, \ldots, r_p, \hat{r}, \check{r}$ is pointed.

PROOF: The result is based on the separation theorem for pointed cones (see Appendix C). As C is pointed, there exists $\lambda \in lin(C)$ such that $\lambda \cdot c > 0$, for all $c \in C$, with $c \neq 0$. In particular, $\lambda \cdot r > 0$ and $\lambda \cdot r_j > 0$, for all $j = 1, \ldots p$. As $\rho \in C^{\perp}$, then $\lambda \cdot \rho = 0$.

We prove that $\lambda \cdot c > 0$, for all $c \in C'$, with $c \neq 0$. Trivially the property is true for each of the generators r_j of C. Besides:

$$
\begin{aligned}
\lambda \cdot \hat{r} &= \lambda \cdot r + \lambda \cdot \rho = \lambda \cdot r > 0 \\
\lambda \cdot \check{r} &= \lambda \cdot r - \lambda \cdot \rho = \lambda \cdot r > 0
\end{aligned}
$$

Therefore, C' is pointed. ∎ 3.3.1

Proposition 3.3.2   Let C be a pointed polyhedral convex cone with generators $r_1, \ldots, r_p$. Let $\rho \in C^{\perp}$ and $r \in C$, with $\rho \neq 0$ and $r \neq 0$. Define $\hat{r} = r + \rho$. Then the cone C' generated by $r_1, \ldots, r_p, \hat{r}, -\rho$ is pointed.

PROOF: The result is based on the separation theorem for pointed cone (see Appendix C). As C is pointed, there exists $\lambda \in lin(C)$ such that $\lambda \cdot c > 0$, for all $c \in C$, with $c \neq 0$. In particular, $\lambda \cdot r > 0$ and $\lambda \cdot r_j > 0$, for all $j = 1, \ldots, p$. As $\rho \in C^{\perp}$, then $\lambda \cdot \rho = 0$.

Let $\lambda' = a\lambda - \rho$, with $a > |\rho|^2/(\lambda \cdot r) > 0$. We prove that $\lambda' \cdot c > 0$ for all $c \in C'$, with $c \neq 0$. In fact, for all $c \in C$, with $c \neq 0$, $\lambda' \cdot c = a\lambda \cdot c - \rho \cdot c = a\lambda \cdot c > 0$. Hence, $\lambda' \cdot r_j > 0$ for all $j = 1, \ldots, p$. Besides:

$$
\begin{aligned}
\lambda' \cdot \hat{r} &= \lambda' \cdot (r + \rho) = a\lambda \cdot (r + \rho) - \rho \cdot (r + \rho) \\
&= a\lambda \cdot r + a\lambda \cdot \rho - \rho \cdot r - \rho \cdot \rho = a\lambda \cdot r - |\rho|^2 > 0 \\
\lambda' \cdot (-\rho) &= -a\lambda \cdot \rho + \rho \cdot \rho = |\rho|^2 > 0
\end{aligned}
$$

Therefore, C' is pointed. ∎ 3.3.2

Note that if C is a dependence cone and λ corresponds to a valid affine timing function for C, the proofs of the above propositions indicate how to determine a vector λ' from λ, which defines a valid affine timing function for the dependence cone C'. In particular, according to Proposition 3.3.1, λ' is equal to λ, while according to Proposition 3.3.2, λ' is equal to $a\lambda - \rho$, with $a > |\rho|^2/(\lambda \cdot r)$.

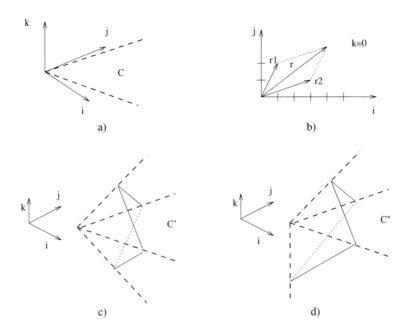

Fig. 3.17. a) Cone C; b) Vector r in C; c) Cone C' d) Cone C''.

Example 3.3.3   Consider the pointed cone C of Fig. 3.17 a), generated by $r_1 = (1, 2, 0)$, $r_2 = (3, 1, 0)$, and the non-null vector $\lambda = (0, 1, 0)$. Note that $\lambda \cdot r_i > 0$, for $i = 1, 2$. Consider the vector $r = r_1 + r_2 = (4, 3, 0)$ in C (see Fig. 3.17 b)) and the vector $\rho = (0, 0, 1)$ in $C^\perp$.

Let us first consider the cone C' (see Fig. 3.17 c)) generated by $r_1, r_2, \hat{r}$ and $\check{r}$, where $\hat{r} = r + \rho = (4, 3, 1)$ and $\check{r} = r - \rho = (4, 3, -1)$. Then C' is pointed and $\lambda \cdot c > 0$ for all $c \in C'$. This fact can be verified by considering the generators of C', i.e., :

$$\begin{aligned}
\lambda \cdot r_1 &= 2 > 0 \\
\lambda \cdot r_2 &= 1 > 0 \\
\lambda \cdot \hat{r} &= 3 > 0 \\
\lambda \cdot \check{r} &= 3 > 0
\end{aligned}$$

Let us now consider the cone C'' (see Fig. 3.17 d)) generated by $r_1, r_2, \hat{r}$ and $-\rho$, where $\hat{r} = r + \rho = (4, 3, 1)$ and $-\rho = (0, 0, -1)$. Then C'' is pointed and any $\lambda' = a\lambda - \rho$, with $a > 1/3$, is such that $\lambda' \cdot c > 0$ for all $c \in C''$. For

Technique	Generators of C	Generators of C'	Proposition
Unif. (3.2.10)	r	r	—
Unif. (3.2.13)	r	$r, \hat{r}, \check{r}$	3.3.1
Par. Unif. (3.2.16)	r	$r, \hat{r}, \check{r}$	3.3.1
Decomp. (3.2.19)	$r, r_1, \ldots, r_m$	$r, r_1, \ldots, r_m$	—
Decomp. (3.2.21)	$r, r_1, \ldots, r_m$	$r, r_1, \ldots, r_m, \hat{r}, \rho$	3.3.2
Decomp. (3.2.23)	$r, r_1, \ldots, r_m$	$r, r_1, \ldots, r_m$	—

Table 3.1. Regularisation and the preservation of affine scheduling.

instance, let $a = 1$ and $\lambda' = \lambda - \rho = (0, 1, -1)$. Then:

$$\lambda' \cdot r_1 = 2 > 0$$
$$\lambda' \cdot r_2 = 1 > 0$$
$$\lambda' \cdot \hat{r} = 2 > 0$$
$$\lambda' \cdot (-\rho) = 1 > 0$$

Note that λ does not have the same property as, for instance, $\lambda \cdot (-\rho) = 0$.

■ 3.3.3

Both the uniformisation and decomposition techniques which we have introduced for integral data dependencies define particular cases to which the above propositions apply. We have summarised the various techniques in Table 3.1. The first column of the table indicates the technique with, in brackets, the reference number of the corresponding formal result. The generators of the embedding dependence cone before (C) and after (C') the application of the technique are given in the second and third columns, respectively. In particular, in such columns $r = -d$ and $r_j = -d_j$, for all $j = 1, \ldots, m$, denote the generators of the corresponding dependence cone, while $\hat{r} = -\hat{d}, \check{r} = -\check{d}$ and $\rho = -\pi$ the new generators introduced by the various techniques. The last column indicates which of the above proposition applies. A line — indicates that no result applies and corresponds to a case in which the dependence cone is not modified by the regularisation technique.

From the above results, it follows that the integral regularisation techniques which we have defined preserve the affine scheduling of a data dependence. In order to guarantee a similar result more globally for the whole specification we need to consider all its data dependencies at the same time.

Fortunately, this is not a major problem. In fact, we may define the dependence cone of a system of equations **S** as the smallest polyhedral convex cone which contains all the dependence cones of **S**. Let $C_{\mathbf{S}}$ denote such a cone. Then, by definition, for each data dependence $\mathcal{DD}$ of **S**, the corresponding cone C is contained in $C_{\mathbf{S}}$. Hence, a choice of π in $(C_{\mathbf{S}})^{\perp}$ guarantees both that $\pi \in C^{\perp}$ and that affine schedulings are preserved for the system **S**.

3.4   Summary

In this chapter we have discussed how integral data dependencies can be defined and made uniform in the context of classic regular array synthesis. The key issue was the relationship established between the integral index mapping and sets of direction vectors of the lattice space. This relationship has allowed us both to relate integral data dependencies to affine data dependencies, and to generalise forms of regularisation techniques from the affine to the integral case.

We formulated the notion of atomic integral data dependence as an integral data dependence with a particular rôle in regularisation. Based on such a notion, we have developed both decomposition and uniformisation techniques. We have summarised those techniques in Table 3.2. In the table, the first column indicates the technique and in brackets the reference number of the corresponding formal result. The second column indicates the basic conditions for its applicability. In particular, d denotes a generator of the integral index mapping, D the domain of the data dependence, and P the polyhedron generated by d and D. The last column indicates whether the application of the technique may require an increase in the number of dimensions of the space.

The techniques which we have developed are amenable to semi-automatic support. Algorithm designer's intervention is required, typically for the choice of the regularisation directions (the generators of the integral index mapping and their ordering) as well as the hyperplane on which pipelining and routing schemes are based (the vector π in the formal propositions). The details of the transformations are, however, accounted for by the techniques. Also, the conditions for the existence and preservation of affine scheduling which we have developed are not just of theoretical relevance, but provide guidelines to the designer in the choice of the regularisation directions.

As affine data dependencies are particular cases of integral data dependencies, several techniques are available for their regularisation, including

Technique	Condition	Extra Dimensions
Unif. (3.2.10)	$d \notin lin(D)$	no
Unif. (3.2.13)	$d \in lin(D)$ $dim(D) < n$	yes
Par. Unif. (3.2.16)	$d \in lin(D)$ $dim(D) < n$	yes
Decomp. (3.2.19)	$d \notin lin(D)$	no
Decomp. (3.2.21)	$d \in lin(D)$ $dim(P) < n$	yes
Decomp. (3.2.23)	known inverse	no

Table 3.2. Summary of regularisation techniques.

those specifically developed in the literature, and those we have developed in this work for the integral case. In general, the designer has to evaluate which technique should be adopted on the basis of the requirements of each specific problem. As a general rule, however, the more specialised a technique is, the simpler and more effective it is likely to be. For instance, integral techniques always introduce control overhead as they are tailored to situations in which convexity needs to be recovered. Indeed, in the affine case convexity comes for free. Also, the removal of affine broadcasts can be achieved very effectively through a technique (known in the literature as *pipelining* [QuVa89]) which selects regularisation directions in the null space of the linear part of the affine transformation defined by an index mapping. A similar technique cannot be adopted for integral mappings, as linearity is not one of their general properties. On the other hand, the regularisation schemes that we have developed have a property of reconfigurability, achieved by the initialisation of the control variables, which is not characteristics of any affine regularisation technique. Our techniques are also, in general, more detailed than the affine techniques presented in [QuVa89] (on which our work in based), which tend to state the applicability conditions without providing all the details for their practical implementation.

Chapter 4

Dynamic Recurrence Equations

Both affine and integral data dependencies are *static* in that they are completely defined when the algorithm is specified. There exist algorithms, however, whose data dependencies do not share this property: for example, their definition may rely upon values which are provided or computed only when the algorithm is executed. We call such data dependencies *dynamic*. In this chapter we aim at introducing the concept of dynamic data dependence in the context of regular array synthesis and identifying classes of dynamic data dependencies which are amenable to a systematic transformation into regular arrays.

The development of synthesis techniques for dynamic problems constitutes an entirely new chapter in regular array design. Researchers in the field have so far avoided consideration of dynamic problems, mainly on the belief that dynamic dependencies and the static topology of regular arrays cannot be reconciled. However, examples of *ad hoc* regular arrays for problems which may be considered as dynamic exist in the literature (one of these problems, that of Gaussian elimination with pivoting, will be considered in Chapter 5). Therefore, it appears that regular array design is feasible at least for restricted classes of dynamic problems.

One of the basic issues we need to address in the synthesis of dynamic problems is the type of dynamic data dependence relation that we want to consider. The notion we adopt in this chapter is that of a data dependence relation which is dynamic with respect to the inputs of the algorithm. More precisely, the data dependence relation between two variables of a specification changes (i.e., different pairs of their instances may be related under the data dependence) at each execution of the algorithm on the basis of the input values provided. In formalising this notion in the context of regular array synthesis, a major difficulty to overcome is the lack, in the classical framework, of some of the necessary basic concepts, primarily a formal notion of input. Hence, in the first part of this chapter we introduce some of these basic notions. In particular, we provide formal definitions of input and indexed variable, in Sections 4.1, and of dynamic index mapping, dynamic data dependence and dependence graph, in Section 4.2.

In the second part of this chapter, we identify classes of dynamic problems which can be reduced systematically to uniform problems. The problems we consider may be regarded as a dynamic generalisation of integral problems. For such problems, regularisation is made possible by the use of control variables which can be dynamically reconfigured at each execution of the algorithm. Reconfigurable control variables enable us to reconcile the existence of dynamic data dependencies in the algorithm with its execution on a regular array of static topology. Namely, the connections between processing elements are not altered, while software control provides for a flexible routing of the data through the network. Sections 4.3, 4.4 and 4.5 deal with dynamic problems and their regularisation. Finally, we draw some conclusions in Section 4.6.

4.1   Inputs and Indexed Variables

As mentioned above, we consider data dependencies which are dynamic with respect to the inputs of a system of equations. For their characterisation we need first to define more precisely what an input is. The basic concepts of regular array synthesis introduced in Chapter 2 do not account for a such a definition, because inputs do not play an essential part in the analysis and transformation of static data dependencies, hence their explicit treatment can be avoided.

By recalling the notation of Chapter 2, we will denote by $\mathcal{CS}$ the computation space and Val the set of data values, containing the undefined value $\perp$.

Given a system of equations $\mathbf{S}$, we define an *input of* $\mathbf{S}$ as an assignment of values to some of the variables of $\mathbf{S}$ at points of the computation space. As may be expected, such an assignment is based on the input equations of $\mathbf{S}$. More precisely:

Definition 4.1.1 *[Inputs]* Let $\mathbf{S}$ be a system of equations with input equations $\mathbf{E}_1, \ldots, \mathbf{E}_r$, where $\mathbf{E}_i = (D_i, U_i, in_i)$, for $i = 1, \ldots, r$. Let $\mathcal{V} = \{U_1, \ldots, U_r\}$ and $D = \bigcup_{i=1}^{r} D_i$. We define an input of $\mathbf{S}$ as a mapping $in : \mathcal{V} \times D \to Val$, where:

$$in(U_i, c) = \begin{cases} in_i(c) & c \in D_i \\ \bot & \text{otherwise} \end{cases}$$

∎ 4.1.1

We denote the set of inputs to $\mathbf{S}$ by $Input_{\mathbf{S}}$.

Example 4.1.2   Consider the following system of equations:

$$\begin{aligned} \mathbf{E}_1 &= (D_1, F, in_F) \\ \mathbf{E}_2 &= (D_2, F, (F, F), +, (\mathcal{I}_1, \mathcal{I}_2)), \end{aligned}$$

with domains $D_1 = \{0, 1\}$, $D_2 = \{2, \ldots, n\}$, for some integer $n > 2$; index mappings $\mathcal{I}_1(i) = i - 1$, $\mathcal{I}_2(i) = i - 2$; and applied functions $+(a, b) = a + b$, $in_F(i) = f_i$, with $f_i \in \mathbf{Z}$. This system of equations generalises that in Example 2.1.2 for the Fibonacci sequence. The difference with respect to Example 2.1.2 is the definition of in_F that, here, is not a constant function.

$\mathbf{E}_1$ is the only input equation of the system. The inputs of $\mathbf{S}$ are functions of the type $in : \{F\} \times D_1 \to \mathbf{Z} \cup \{\bot\}$, where:

$$in(F, i) = \begin{cases} in_F(i) & i \in D_1 \\ \bot & \text{otherwise} \end{cases}$$

∎ 4.1.2

The definition of inputs above allows us to define more precisely the indexed variables of a system of equations. In particular, while in Chapter 2 we informally associated indexed variables with the (tabulated) functions computed by the algorithm, we can now define an indexed variable as a mapping which associates a (tabulated) function with each input of the algorithm. The definition is the following:

Definition 4.1.3 *[Indexed Variable]* Let $\mathbf{S}$ be a system of equations and $Input_{\mathbf{S}}$ its set of inputs. A variable V of $\mathbf{S}$, with definition domain $Def D_V$, is a mapping from $Input_{\mathbf{S}}$ to $[Def D_V \to Val]$.   ∎ 4.1.3

Example 4.1.4   Let us consider the system of equations in Example 4.1.2. For $in_F(i) = 1$ on D_1, variable F represents the first $n + 1$ entries of the Fibonacci sequence. For $in_F(i) = 0$, F is the zero function. For $in_F(i) = 2$, F represents the first $n + 1$ entries of the sequence $2, 2, 4, 6, 10, \ldots$. Indeed infinitely many functions may be obtained.   ■ 4.1.4

In the following, given a variable V, we adopt the notation $V[\_]$, where $[\_]$ is a place holder for inputs. Then, for all $in \in Input_S$, $V[in]$ defines a function from $DefD_V$ to Val.

The following properties characterise indexed variables which assume integer values. They will be used in the second part of this chapter for the definition of a class of dynamic data dependencies in the context of Euclidean synthesis methods. They are introduced here for ease of presentation.

Definition 4.1.5 *[Integer-valued Indexed Variable]* Let **S** be a system of equations, $Input_S$ its set of inputs, and V a variable of **S** with definition domain $DefD_V$. V is integer-valued if V defines a mapping from $Input_S$ to $[DefD_V \to \mathbf{Z}]$. In addition, let $D \subseteq DefD_V$. Then V is:

- *non-negative* over D, if for all $in \in Input_S$, $V[in](c) \geq 0$ for all $c \in D$;

- *bounded (above)* over D, if there exists $m \in \mathbf{Z}$, such that for all $in \in Input_S$, $V[in](c) \leq m$ for all $c \in D$.

                                    ■ 4.1.5

We assume an integer-valued indexed variable V equal to 0 outside its definition domain, i.e., for all $in \in Input_S$, $V[in](c) = 0$ for $c \notin DefD_V$. We make this assumption as it simplifies the definitions of the regularisation techniques of the following sections. This assumption replaces that made in Chapter 2 that a variable is undefined (equal to $\perp$) outside its definition domain. This implies a slight change in the semantics as we cannot distinguish when the variable is defined and equal to 0 from when it is undefined.

4.1.1   Implicit Quantification

As index variables are defined in terms of inputs, all definitions which involve indexed variables are implicitly universally quantified over those inputs. For instance, a recurrence equation

$$\mathbf{E} = (D, U, (V_1, \ldots, V_m), f, (\mathcal{I}_1, \ldots, \mathcal{I}_m))$$

in a system $\mathbf{S}$ can be seen as a short-hand for the expression: for all $in \in$ $Input_{\mathbf{S}}$,

$$\mathbf{E}[in] = (D, U[in], (V_1[in], \ldots, V_m[in]), f, (\mathcal{I}_1, \ldots, \mathcal{I}_m)).$$

This implicit quantification is not usually addressed in the synthesis methods (indeed, we have ignored it in the previous chapters) as only static elements of the system are manipulated. In particular, the emphasis is on the analysis and transformation of data dependence relations between computation points, and those relations are defined as static, as both computation points and index mappings do not depend on the inputs of a system of equations. However, the shift from static to dynamic data dependencies addressed in the following sections, will require us to consider the inputs of the system explicitly. In the following, we will retain the implicit quantification whenever possible.

4.2   Dynamic Data Dependencies

Our concept of dynamic data dependencies is based on the idea of dynamic index mapping as an index mapping which is dynamic with respect to the inputs of an algorithm. That an index mapping becomes dynamic may appear as a simple extension of the static notion. It has, however, far reaching repercussion on the whole formalism, as index mappings are among the basic elements of synthesis methods. In particular, data dependence relations and their graphical representation as dependence graphs are affected, in that a new dependence relation and graph are obtained for each input. Formally, we define a dynamic index mapping as follows:

Definition 4.2.1 *[Dynamic Index Mapping]* Let $\mathbf{S}$ be a system of equations and $Input_{\mathbf{S}}$ its set of inputs. A dynamic index mapping $\mathcal{I}$ is a mapping from $Input_{\mathbf{S}}$ to $[\mathcal{CS} \rightarrow \mathcal{CS}]$.   ■ 4.2.1

Definition 4.2.1 implies that for all $in \in Input_{\mathbf{S}}$, $\mathcal{I}[in]$ is an index mapping. Note that the definition assumes the existence of a system of equations upon whose inputs the index mapping is defined. For convenience, in the following we will omit the explicit reference to this system, and will denote the set of its inputs simply by $Input$, without decoration.

Note that a (static) index mapping can be interpreted as a dynamic index mapping with a singleton set as its range, i.e., the dynamic index mapping always associates the same (static) index mapping with all the inputs to the system.

Example 4.2.2   Consider the following single assignment code segment:

```
for i := 1 to 3 do
  read(Y(i));
for i := 4 to n do
  begin
    read(G(i));
    read(X(i));
    I(i) := X(i) − (G(i) mod 4);
      Y(i) := sqr(Y(i − I(i)));
  end;
```

The code corresponds to the system **S** of equations:

$$
\begin{aligned}
\mathbf{E}_1 &= (D_1, Y, in_Y) \\
\mathbf{E}_2 &= (D_2, G, in_G) \\
\mathbf{E}_3 &= (D_2, X, in_X) \\
\mathbf{E}_4 &= (D_2, I, (X, G), h, (\mathcal{I}_1, \mathcal{I}_1)) \\
\mathbf{E}_5 &= (D_2, Y, Y, sqr, \mathcal{I}_2)
\end{aligned}
$$

with domains $D_1 = \{0, \ldots, 3\}$ and $D_2 = \{4, \ldots, n\}$, index mappings $\mathcal{I}_1(i) = i$, $\mathcal{I}_2(i) = i - I(i)$, and applied functions $h(a, b) = a - (b \bmod 4)$, $sqr(a) = a^2$, $in_Y(i) = y_i$, with $y_i \in \mathbf{R}$, $in_G(i) = g_i$, with $g_i \in \mathbf{Z}$, and $in_X(i) = i$.

The mapping $\mathcal{I}_2(i) = i - I(i)$ is a dynamic index mapping. In fact, $\mathcal{I}_2$ is defined in terms of the indexed variable I that, by definition, depends on the inputs of **S**.   ∎ 4.2.2

Dynamic data dependencies and data dependence graphs can be defined on the basis of dynamic index mappings. The main difference from the static case is that for each input a new relation and corresponding graph are defined, based on the corresponding index mapping. The reader may compare the definitions below with those in Chapter 2.

Definition 4.2.3 *[Dynamic Data Dependence]* Let $\mathcal{I}$ be a dynamic index mapping. A dynamic data dependence $\mathcal{DD}$ is a 4-tuple $(D, U, V, \mathcal{I})$, such that for all $in \in Input$, $\mathcal{DD}[in] = (D, U, V, \mathcal{I}[in])$ is a data dependence.   ∎ 4.2.3

Definition 4.2.4 *[Dynamic Data Dependence Graph]* Let $\mathcal{DD} = (D, U, V, \mathcal{I})$ be a dynamic data dependence. Its dynamic data dependence graph $\mathcal{DDG}$ is a 2-tuple $(\mathcal{N}, \mathcal{A})$, such that for all $in \in Input$, $(\mathcal{N}[in], \mathcal{A}[in])$ is the data dependence graph defined by:

- $\mathcal{N}[in] = D \cup \mathcal{I}[in](D)$; and

- $\mathcal{A}[in] = \{(\mathcal{I}[in](d), d) \mid d \in D\}$.

                                                                    ■ 4.2.4

Example 4.2.5   Consider the system **S** of Example 4.2.2. Its data dependencies are:

$$\begin{aligned}
\mathcal{DD}_1 &= (D_2, I, X, \mathcal{I}_1) \\
\mathcal{DD}_2 &= (D_2, I, G, \mathcal{I}_1) \\
\mathcal{DD}_3 &= (D_2, Y, Y, \mathcal{I}_2)
\end{aligned}$$

with $\mathcal{I}_1(i) = i$ and $\mathcal{I}_2(i) = i - I(i)$. The data dependencies $\mathcal{DD}_1$ and $\mathcal{DD}_2$ are (static) uniform, while $\mathcal{DD}_3$ is dynamic. The data dependence graph defined by $\mathcal{DD}_3$ is also dynamic. Fig. 4.1 a) and b) illustrate the dynamic data dependence graph for two inputs to **S** and for $n = 11$. The related tables give the corresponding values of the variables of the system on their domains.

                                                                    ■ 4.2.5

4.3   Dynamic Data Dependencies in Euclidean Synthesis

In this section we introduce a class of dynamic data dependencies which are amenable to systematic transformation into regular arrays in the context of Euclidean synthesis methods .

The benefits of adopting Euclidean geometry and linear algebra as the mathematical framework for synthesis methods, both from a theoretical and applicative point of view, have already been discussed at length in Chapter 2. We have also already mentioned the necessity of providing an explicit syntactic characterisation for the formal manipulation of algorithm specifications in such a context (see Section 2.2.1).

i	0	1	2	3	4	5	6	7	8	9	10	11
$G(i)$	⊥	⊥	⊥	⊥	4	5	6	7	8	9	10	11
$X(i)$	⊥	⊥	⊥	⊥	4	5	6	7	8	9	10	11
$I(i)$	0	0	0	0	4	4	4	4	8	8	8	8
$i - I(i)$	0	1	2	3	0	1	2	3	0	1	2	3
$Y(i)$	y_0	y_1	y_2	y_3	y_0^2	y_1^2	y_2^2	y_3^2	y_0^2	y_1^2	y_2^2	y_3^2

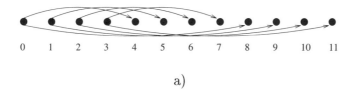

a)

i	0	1	2	3	4	5	6	7	8	9	10	11
$G(i)$	⊥	⊥	⊥	⊥	2	3	5	2	7	11	9	17
$X(i)$	⊥	⊥	⊥	⊥	4	5	6	7	8	9	10	11
$I(i)$	0	0	0	0	2	2	5	5	5	6	9	10
$i - I(i)$	0	1	2	3	2	3	1	2	3	3	1	1
$Y(i)$	y_0	y_1	y_2	y_3	y_2^2	y_3^2	y_1^2	y_2^2	y_3^2	y_3^2	y_1^2	y_1^2

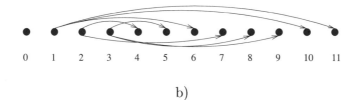

b)

Fig. 4.1. Values of the variables of **S** for two inputs to the system and relative data dependence graphs.

In the following we will provide a syntactic characterisation of a class of dynamic data dependencies, which we see as a natural dynamic generalisation of integral data dependencies. The generalisation is provided by replacing integer functions with integer-valued variables as the coefficients of the index mappings. Such a replacement introduces a form of dynamicity which depends on the inputs of the system of equations.

The main benefit of this approach is that regularisation techniques can be developed from those of integral data dependencies. In particular, we will show that the reconfigurability characteristics of the integral routing scheme that we have defined, provides the necessary flexibility for the treatment of this particular type of dynamic problems. There is, however, a restriction to their applicability, which is formally captured by the notion of separability between variables. Separability indicates that all computations necessary to resolve the dynamic data dependence relations can be separated and carried out prior to any other computation of the algorithm. Intuitively, if an algorithm is specified as a sequence of nested for-loops, separability allows us to reorganise the nested for-loops so that the control variables which define the dynamic routing of the data are computed initially, while all remaining nested for-loops follow. This type of restructuring of the computations implies that a further updating of the control signals is not possible. That is, once a dynamic data dependence relation has been established, it becomes static for the rest of the execution of the algorithm. From a mathematical point of view, the reorganisation of the computations is achieved through geometric translations of computation domains and a redefinition of the relative equations. That separability constitute a restriction for the regularisation of dynamic problems will become clear also from the case studies of Chapter 5.

For presentation purposes, the remainder of the chapter is structured similar to Chapter 3 for integral recurrences. Such an organisation facilitates the comparison between the two approaches. The main differences to keep in mind concern the notion of separability and the corresponding reorganisation of the computations. Other minor differences will be highlighted in the discussion.

4.3.1   Finitely Generated Index Mapping

A natural dynamic generalisation of an integral index mapping is the following (the reader may compare this definition with Definition 3.1.1 in Chapter 3), which replaces integer functions with integer-valued variables as the coefficients of the index mapping:

Definition 4.3.1 *[Finitely Generated Index Mapping]* Let $\mathcal{I}$ be a dynamic index mapping. $\mathcal{I}$ is finitely generated if for all $in \in Input$ and for all $z \in \mathbf{Z}^n$, $\mathcal{I}[in](z) = z + \sum_{j=1}^{m} G_j[in](z)d_j$, where, for $j = 1, \ldots, m$, G_j is an integer-valued indexed variable and d_j is a non-null vector in $\mathbf{Z}^n$. In addition, $\mathcal{I}$ is atomic finitely generated if $m = 1$. ■ 4.3.1

Variables G_j are called the coefficients of the index mapping and vectors d_j its generators. Following the convention in Section 4.1.1, in the following, when possible, we will express such an index mapping $\mathcal{I}$ more succinctly as $\mathcal{I}(z) = z + \sum_{j=1}^{m} G_j(z)d_j$, i.e., we will leave the quantification over the inputs implicit.

Example 4.3.2 Consider the system of equations in Example 4.2.2. The index mapping $\mathcal{I}_2$, defined as $\mathcal{I}_2(i) = i - I(i)$, is finitely generated and, in particular, is atomic. The generator is the vector (-1) with coefficient the integer-valued variable I. ■ 4.3.2

Similar to the integral case we may define:

Definition 4.3.3 *[Finitely Generated Dependence Mapping]* Let $\mathcal{I}$ be a finitely generated index mapping. Its dependence mapping $\Theta_{\mathcal{I}}$ is such that, for all $in \in Input$ and for all $z \in \mathbf{Z}^n$, $\Theta_{\mathcal{I}}[in](z) = z - \mathcal{I}[in](z)$. ■ 4.3.3

If $\mathcal{I}$ is expressed as $\mathcal{I}(z) = z + \sum_{j=1}^{m} G_j(z)d_j$, its dependence mapping can be expressed as $\Theta_{\mathcal{I}}(z) = -\sum_{j=1}^{m} G_j(z)d_j = \sum_{j=1}^{m} G_j(z)(-d_j)$. We call the vectors $-d_j$ the generators of $\Theta_{\mathcal{I}}$ and the variables G_j its coefficients.

4.3.2 Finitely Generated Data Dependence

Finitely generated index mappings characterise finitely generated dynamic data dependencies:

Definition 4.3.4 *[Finitely Generated Data Dependence]* Let $\mathcal{DD} = (D, U, V, \mathcal{I})$ be a dynamic data dependence. $\mathcal{DD}$ is finitely generated if $\mathcal{I}$ is a finitely generated index mapping. In addition, $\mathcal{DD}$ is atomic finitely generated if $\mathcal{I}$ is atomic finitely generated. ■ 4.3.4

The dependence domain and dependence cone of a finitely generated data dependence can be defined accordingly:

Definition 4.3.5 *[Finitely Generated Dependence Domain]* Let $\mathcal{DD} = (D, U, V, \mathcal{I})$ be a finitely generated data dependence and $\Theta_{\mathcal{I}}$ the dependence mapping defined by $\mathcal{I}$. The dependence domain $\Omega_{\mathcal{I}}$ of $\mathcal{DD}$ is such that for all $in \in Input$, $\Omega_{\mathcal{I}}[in] = \Theta_{\mathcal{I}}[in](D)$. ∎ 4.3.5

Definition 4.3.6 *[Finitely Generated Dependence Cone]* Let $\mathcal{DD} = (D, U, V, \mathcal{I})$ be a finitely generated data dependence with dependence domain $\Omega_{\mathcal{I}}$. The dependence cone $\Theta_{\mathcal{I}}^*$ of $\mathcal{DD}$ is such that for all $in \in Input$, $\Theta_{\mathcal{I}}^*[in] = cone(\Omega_{\mathcal{I}}[in])$. ∎ 4.3.6

Example 4.3.7   Consider the finitely generated data dependence $\mathcal{DD} = (D, U, V, \mathcal{I})$ with index mapping $\mathcal{I}(i, j) = (i, j) + G_1(i, j)(1, 1) + G_2(i, j)(1, 0)$ and domain $D = \{(i, j) \mid 1 \le i, j, \le n\}$, for some $n \in \mathbf{N}$. Assume that the coefficients of $\mathcal{I}$ are defined by the input equations:

$$
\begin{aligned}
\mathbf{E}_1 &= (D, G_1, in_{G_1}) \\
\mathbf{E}_2 &= (D, G_2, in_{G_2})
\end{aligned}
$$

with $in_{G_1}(i, j) \in \mathbf{Z}$ and $in_{G_2}(i, j) = i + j$, for all $(i, j) \in D$.

The points of the dependence domains and the dependence cones of $\mathcal{DD}$ for two inputs are illustrated in Fig. 4.2 (for $n = 3$). The values of the variables and of the index and dependence mappings for each point of D are given in the related tables. ∎ 4.3.7

4.3.3   Dependence Cone and Pointedness

From a previous discussion (see Sections 2.2.6), we know that pointed dependence cone guarantee the existence of an affine scheduling. For an integral data dependencies this property was exploited as a guideline in the choice of the generators of the index mapping, and sufficient conditions were given for an integral dependence cone to be pointed and integral regularisation techniques to realise transformations of pointed dependence cones (see Sections 3.2.1 and 3.3).

In this section we investigate the possibility of formulating similar conditions for finitely generated data dependencies. The main difficulty here is that a dynamic dependence cone actually defines a family of dependence cones, one for each input, and that each such cone is known only when the

(i,j)	$G_1(i,j)$	$G_2(i,j)$	$\mathcal{I}(i,j)$	$\Theta_{\mathcal{I}}(i,j)$
$(1,1)$	3	2	$(6,4)$	(-5,-3)
$(1,2)$	2	3	$(6,4)$	(-5,-2)
$(1,3)$	5	4	$(10,8)$	(-9,-5)
$(2,1)$	1	3	$(6,2)$	(-4,-1)
$(2,2)$	1	4	$(7,3)$	(-5,-1)
$(2,3)$	2	5	$(9,5)$	(-7,-2)
$(3,1)$	3	4	$(10,4)$	(-7,-3)
$(3,2)$	2	5	$(10,4)$	(-7,-2)
$(3,3)$	1	6	$(10,4)$	(-7,-1)

(i,j)	$G_1(i,j)$	$G_2(i,j)$	$\mathcal{I}(i,j)$	$\Theta_{\mathcal{I}}(i,j)$
$(1,1)$	2	2	$(5,3)$	(-4,-2)
$(1,2)$	3	3	$(7,5)$	(-6,-3)
$(1,3)$	2	4	$(7,5)$	(-6,-2)
$(2,1)$	4	3	$(9,5)$	(-7,-4)
$(2,2)$	1	4	$(7,3)$	(-5,-1)
$(2,3)$	1	5	$(8,4)$	(-6,-1)
$(3,1)$	5	4	$(11,6)$	(-9,-5)
$(3,2)$	1	5	$(9,3)$	(-6,-1)
$(3,3)$	2	6	$(11,5)$	(-8,-2)

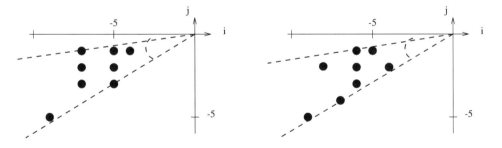

Fig. 4.2. Dependence domains and cones for two inputs of G_1 and G_2.

algorithm is executed. Hence, in general, we cannot decide at compile time
if all the cones of the family will be pointed. However, we may be able to
guarantee that this is the case if we can make a number of assumptions on
its finitely generated index mapping. The result is contained in the following
proposition:

Proposition 4.3.8   Let $\mathcal{DD} = (D, U, V, \mathcal{I})$ be a finitely generated data
dependence with index mapping $\mathcal{I}(z) = z + \sum_{i=1}^{m} G_j(z) d_j$. If:

- the coefficients G_j are non-negative over D; and

- the embedding dependence cone $C = cone(\{-d_1, \ldots, -d_m\})$ is pointed,

then, for all $in \in Input$, $\Theta^*[in]$ is pointed and contained in C.

PROOF: The result follows from the definition of C. In fact, by definition C
        contains all points c in $\mathbf{Z}^n$ such that $c = \sum_{j=1}^{m} a_j(-d_j)$, with $a_j \geq 0$.
        For all inputs in, $\Omega_{\mathcal{I}}[in]$ is defined as the set $\{\sum_{j=1}^{m} G_j[in](z)(-d_j) \mid$
        $z \in D\}$. If for all j, $G_j[in](z) \geq 0$ then $\Omega_{\mathcal{I}}[in] \subset C$ and $\Theta_{\mathcal{I}}^*[in] \subseteq$
        C. Furthermore, because of the inclusion $\Theta_{\mathcal{I}}^*[in] \subseteq C$, C pointed
        implies $\Theta_{\mathcal{I}}^*[in]$ pointed.   ∎ 4.3.8

From Proposition 4.3.8, it follows that if λ is a non-null vector in $\mathbf{Z}^n$,
such that $\lambda \cdot z > 0$ for each element z of C, then $\lambda \cdot z > 0$ for all $z \in \Theta_{\mathcal{I}}^*[in]$
and for all inputs in. Therefore any affine timing function defined by λ is
valid for all data dependence cones $\Theta_{\mathcal{I}}^*[in]$.

In the following we always assume that the conditions of Proposition 4.3.8
are satisfied. Note that if this assumption is not fulfilled, we can guarantee neither the existence of a timing function for all inputs, nor that our
regularisation techniques will preserve affine scheduling.

Example 4.3.9   Consider the data dependence of Example 4.3.7 and
its dependence cones in Fig. 4.2 for two inputs of G_1 and G_2. The cone
$C = cone(\{(-1, -1), (-1, 0)\})$, illustrated in Fig. 4.3, is pointed and contains
both dependence cones.   ∎ 4.3.9

4.3.4   Finitely Generated *vs.* Integral Recurrences

Given a finitely generated data dependence $\mathcal{DD}$, for all $in \in Input$, $\mathcal{DD}[in]$
defines an integral data dependence. Vice versa, any integral data depen-
dence may be regarded as a finitely generated data dependence. This can

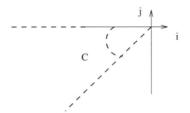

Fig. 4.3. Cone C.

be achieved, for instance, by replacing the coefficients of its index mapping with integer-valued variables, as explained below. Let $\mathcal{DD} = (D, U, V, \mathcal{I})$ be an integral data dependence, with index mapping $\mathcal{I}(z) = z + \sum_{i=1}^{m} g_j(z)d_j$. Then $\mathcal{DD}$ can be transformed in a dynamic data dependence by redefining its index mapping as $\mathcal{I}(z) = z + \sum_{i=1}^{m} G_j(z)d_j$, where each coefficient G_j, for $i = 1, \ldots, m$, is an integer-valued variable defined by the input equation $\mathbf{E}_j = (D, G_j, g_j)$. Note that this transformation is not just a simple syntactic manipulation of the index mapping (as, for instance, that we used to demonstrate that an affine index mapping is also integral – see Section 3.1.1). The change from static to dynamic is a semantic transformation, as the new interpretation modifies the possible models of the specification. Also, the transformation is not unique. Here we have chosen the simplest transformation we could think of. The reader may experiment other ways of interpreting an integral data dependence as a dynamic data dependence.

Note that, although theoretically an interpretation of a static data dependence as a dynamic data dependence is possible, from a practical point of view there is little convenience in doing so, as, in general, dynamic data dependencies require more complex (and often less efficient) forms of regularisation.

By recalling the taxonomy in Section 3.1.1, we may represent the relations between the various classes of recurrences which we have considered in this work, as illustrated in Fig. 4.4. The abbreviations used in the figure are: UREs, for uniform recurrences; AREs, for affine; AIREs and IREs, for atomic integral and integral recurrences; and FGREs, for finitely generated recurrences.

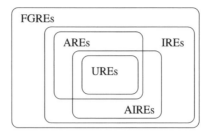

Fig. 4.4. Taxonomy of classes of recurrence equations.

4.3.5   Extended Dependence Graph

By introducing variables in the index mappings, we generate extra dependencies among the variables of a system of equations. In fact, given a finitely generated data dependence $\mathcal{DD} = (D, U, V, \mathcal{I})$ with index mapping $\mathcal{I}(z) = z + \sum_{j=1}^{m} G_j(z)d_j$, for all $z \in \mathbf{Z}^n$, the computation of U depends on the values of G_j as well as the values of V. Let us define the set of variables of $\mathcal{I}$ as $Var_{\mathcal{I}} = \{G_1, \ldots, G_m\}$. For a static index mapping $\mathcal{I}$, which is no interpreted as dynamic, we assume $Var_{\mathcal{I}} = \emptyset$. We define an extended version of the reduced dependence graph of a system of equation as follows:

Definition 4.3.10 *Extended Dependence Graph]* Let $\mathbf{S}$ be a system of equations. Define its extended dependence graph $\mathcal{EDG}$ as the graph $(\mathcal{N}, \mathcal{A})$, where:

- $\mathcal{N} = Var_{\mathbf{S}}$; and

- $\mathcal{A} = \{(U, V) \mid \exists D, \mathcal{I} \text{ such that } (D, U, V, \mathcal{I}) \in \mathcal{DD}_{\mathbf{S}}\}$
  $\cup \{(U, G) \mid \exists D, V, \mathcal{I} \text{ such that } (D, U, V, \mathcal{I}) \in \mathcal{DD}_{\mathbf{S}} \text{ and } G \in Var_{\mathcal{I}}\}$.
  ■ 4.3.10

If no finitely generated index mapping is present in the system, its extended dependence graph coincides with its reduced dependence graph.

Given a system of equations $\mathbf{S}$ and a variable V of $\mathbf{S}$, we can define the subsystem of equations defining V, based on the notion of extended dependence graph. In particular, let $V^{\mathcal{A}}$ denote the set of the nodes of the graph which are accessible from V. Intuitively, this is the set of nodes which can be reached with a path through the graph which starts at V; a formal definition is given in Appendix B. Then the subsystem of equations defining V is $Def\mathbf{S}_V = \cup_{U \in V^{\mathcal{A}}} Def\mathbf{E}_U$. Similarly, for a set of variables $\mathbf{V}$, the subsystem of equations defining $\mathbf{V}$ is $Def\mathbf{S}_{\mathbf{V}} = \cup_{V \in \mathbf{V}} Def\mathbf{S}_V$.

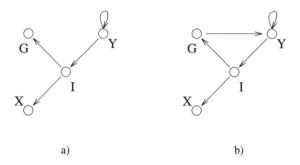

Fig. 4.5. Extended dependence graph of: a) **S**; b) **S**′.

4.3.6 Separability

Separability is a property defined between the variables of a system of equations which allows the identification of separable sets of computations. The notion of separability is based on the dependence relations existing among variables, instead of their instances, i.e., its formulation is based on the extended dependence graph of the system. The definition of separability relies upon the strong connectivity relation of graphs (see Appendix B for a formal definition). We define:

Definition 4.3.11 *[Separability]* Let **S** be a system of equations, $\mathcal{EDG}$ its extended dependence graph and $U, V \in Var_{\mathbf{S}}$. U and V are separable in **S** if and only if U and V are not strongly connected in $\mathcal{EDG}$.   ∎ 4.3.11

Example 4.3.12   Consider the system of equations of Example 4.2.2. Its extended dependence graph is given in Fig. 4.5 a). The equivalence classes of the nodes of the graph under strong connectivity are $\{[G], [Y], [I], [X]\}$. Therefore, all variables are pairwise separable in **S**.

∎ 4.3.12

In the following, given a finitely generated data dependence $\mathcal{DD} = (D, U, V, \mathcal{I})$, we will be interested in the separability of variable U from the coefficients of the index mapping $\mathcal{I}$. If these variables are separable, it will be possible to reorganise their computations in the lattice space and to define, similar to the integral case, a uniform and acyclic routing of the data which substitutes the data dependence $\mathcal{DD}$. Note that when an integral data dependence is regarded as finitely generated (with the interpretation of Section 4.3.4), this separation is always possible, as the coefficients of its index mapping are defined by input equations.

We will not be able to localise finitely generated data dependencies for which this separation is not possible. One such case is illustrated in the following example. The separability of the variables (or the lack of it) is the main restriction to the application of the regularisation techniques developed in this chapter.

Example 4.3.13   Consider the following system of equations $\mathbf{S}'$, obtained by modifying that in Example 4.2.2:

$$
\begin{aligned}
\mathbf{E}_1 &= (D_1, Y, in_Y) \\
\mathbf{E}_2 &= (D_2, G, Y, mod_4, \mathcal{I}_1) \\
\mathbf{E}_3 &= (D_2, X, in_X) \\
\mathbf{E}_4 &= (D_2, I, (X, G), sub, (\mathcal{I}_2, \mathcal{I}_2)) \\
\mathbf{E}_5 &= (D_2, Y, Y, inc, \mathcal{I}_3)
\end{aligned}
$$

with domains $D_1 = \{0, \ldots, 3\}$, $D_2 = \{4, \ldots, n\}$, index mappings $\mathcal{I}_1(i) = i - 1$, $\mathcal{I}_2(i) = i$, $\mathcal{I}_3(i) = i - I(i)$, and applied functions $mod_4(a) = a \bmod 4$, $sub(a, b) = a - b$, $inc(a) = a + 1$, and, on their domains, $in_X(i) = i$ and $in_Y(i) = y_i \in \mathbf{Z}$.

The extended dependence graph of the system is given in Fig. 4.5 b), and the corresponding strongly connected equivalence classes are $\{[G, Y, I], [X]\}$. As Y and I belong to the same equivalence class, they are not separable in $\mathbf{S}'$.   ■ 4.3.13

4.4   Regularisation

The regularisation of finitely generated data dependencies aims at replacing a dynamic specification with a static uniform specification. Once this specification is obtained, classical mapping techniques apply for the derivation of an array design. The transformation from dynamic to static relies upon the definition of control variables which allow us to define a reconfigurable routing scheme for the data without having to alter the topology of the array design.

Regularisation techniques for finitely generated data dependencies extend those for integral data dependencies and have the form of uniformisation and decomposition. The main difference from the integral case stems from the necessity of redefining subsystems of computations of the specification, due to the definition of finitely generated index mappings, in which the coefficients are variables of the system. This characteristics of finitely generated index mappings implies that the corresponding dependence relations can be established only once such coefficients have been computed by the algorithm. Reorganising the computations guarantees that the coefficients are computed prior to their use. The way the computations are reorganised in the computation space of the algorithm is critical for the synchronisation of the data and the possibility of providing affine scheduling. In particular, as an effect of the transformation, new uniform dependence relations are introduced corresponding to acyclic data dependence graphs.

4.4.1 Translation of a System of Equations

We call *translation of a system of equations* the technique which allows us to reorganise the computations of a specification in its computation space. As the name indicates, the technique is based on a geometric translation, that is a simple type of affine transformation.

Systems of equations are translated in the lattice space by translating their computation domains along some direction vector of the space. The technique is defined here with respect to an arbitrary direction vector. In the regularisation techniques of the following section, such a direction vector will correspond to one which is subsequently used for the uniform propagation of the data.

Technically, the translation of a system of equations is a combination of the translation of its domains together with a composition of index mappings which replicates on the translated domains the same data dependence relations of the original domains. In particular, if a computation point x in the original domain is data dependent on a point y under an index mapping $\mathcal{I}$, a similar data dependence relation is established between the translated images of x and y.

As a translation tr is just a mapping in $\mathbf{Z}^n$, the translation of a system of equations can be defined by "lifting" tr from computation points to all the components of the system which are affected by the translation. This lifting is defined in Definition 4.4.1.

Usually we will have to combine a translation of the equations with a renaming of their variables. In fact, in general, the original system of equations is not removed from the algorithm specification, as there might exist other data dependencies in the specification relying on the same equations (this is a side-effect of applying regularisation techniques locally to each data dependence). Indeed all redundant equations may be eliminated once all the necessary regularisations have been completed.

Definition 4.4.1 *[Liftings]* Let $tr : \mathbf{Z}^n \to \mathbf{Z}^n$ be a translation and tr^{-1} denote its inverse. We define the following liftings of tr:

- Let $D \subseteq \mathbf{Z}^n$. Define $tr(D) = \{tr(d) \mid d \in D\}$;

- Let $\mathcal{I} \in [Input \to [\mathbf{Z}^n \to \mathbf{Z}^n]]$. Define $tr(\mathcal{I})$ such that for all $in \in Input$, $tr(\mathcal{I})[in] = tr \circ \mathcal{I}[in] \circ tr^{-1}$;

- Let $\mathcal{IM} \in [Input \to [\mathbf{Z}^n \to \mathbf{Z}^n]]^m$. Define $tr(\mathcal{IM}) = (tr(pr_i(\mathcal{IM})), \dots, tr(pr_m(\mathcal{IM})))$;

- Let $\mathbf{E} = (D_{\mathbf{E}}, {}^\bullet\mathbf{E}, \mathbf{E}^\bullet, f_{\mathbf{E}}, \mathcal{IM}_{\mathbf{E}})$. Define $tr(\mathbf{E}) = (tr(D_{\mathbf{E}}), {}^\bullet\mathbf{E}, \mathbf{E}^\bullet, f_{\mathbf{E}}, tr(\mathcal{IM}_{\mathbf{E}}))$;

- Let $\mathbf{E} = (D_{\mathbf{E}}, {}^\bullet\mathbf{E}, f_{\mathbf{E}})$. Define $tr(\mathbf{E}) = (tr(D_{\mathbf{E}}), {}^\bullet\mathbf{E}, f_{\mathbf{E}} \circ tr^{-1})$;

- Let $\mathbf{S} = \{\mathbf{E}_1, \dots, \mathbf{E}_s\}$. Define $tr(\mathbf{S}) = \{tr(\mathbf{E}_1), \dots, tr(\mathbf{E}_s)\}$.

Let $ren : Var \to Var$ be a variable renaming. We define the following liftings of r:

- Let $\mathbf{V} = (V_1, \dots, V_m) \in Var^m$. Define $ren(\mathbf{V}) = (ren(V_1), \dots, ren(V_m))$.

- Let $\mathbf{E} = (D_{\mathbf{E}}, {}^\bullet\mathbf{E}, \mathbf{E}^\bullet, f_{\mathbf{E}}, \mathcal{IM}_{\mathbf{E}})$. Define $ren(\mathbf{E}) = (D_{\mathbf{E}}, ren({}^\bullet\mathbf{E}), ren(\mathbf{E}^\bullet), f_{\mathbf{E}}, \mathcal{IM}_{\mathbf{E}})$;

- Let $\mathbf{E} = (D_{\mathbf{E}}, {}^\bullet\mathbf{E}, f_{\mathbf{E}})$. Define $ren(\mathbf{E}) = (D_{\mathbf{E}}, ren({}^\bullet\mathbf{E}), f_{\mathbf{E}})$;

- Let $\mathbf{S} = \{\mathbf{E}_1, \dots, \mathbf{E}_s\}$. Define $ren(\mathbf{S}) = \{ren(\mathbf{E}_1), \dots, ren(\mathbf{E}_s)\}$.
                                                                        ■ 4.4.1

By combining the renaming of the variables of a system of equations $\mathbf{S}$ with a translation of its equations, we obtain a new system of equations, which is an image of $\mathbf{S}$ with renamed variables and computations positioned differently from the computations of $\mathbf{S}$ in the lattice space.

Definition 4.4.2 *[Translated Image of a System of Equations]* Let $\mathbf{S}$ be a system of equations. Let $tr : \mathbf{Z}^n \to \mathbf{Z}^n$ be a translation and $ren : Var \to Var$ an injective renaming of the variables of $\mathbf{S}$ with new names. We define the translated image of $\mathbf{S}$ according to tr and ren as $\mathbf{S}^{tr,ren} = ren \circ tr(\mathbf{S})$.
                                                                        ■ 4.4.2

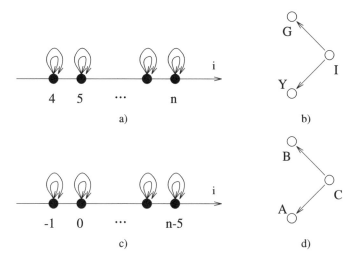

Fig. 4.6. System $(Def\mathbf{S}_I)$: a) data dependence graph; b) reduced dependence graph. System $(Def\mathbf{S}_I)^{tr,ren}$: c) data dependence graph; d) reduced dependence graph.

Example 4.4.3   Consider the system $\mathbf{S}$ of equations of Example 4.2.2, and its sub-system $Def\mathbf{S}_I = \{\mathbf{E}_2, \mathbf{E}_3, \mathbf{E}_4\}$, defining variable I. The data dependence graph and reduced dependence graphs of $Def\mathbf{S}_I$ are illustrated in Fig. 4.6 a) and b), respectively. Let tr be the translation $tr(i) = i - 5$ and ren a renaming such that $ren(I) = C, ren(G) = B$ and $ren(X) = A$. Then $(Def\mathbf{S}_I)^{tr,ren}$ is the following system of equations:

$$\begin{aligned}
\mathbf{E}_2' &= (D_2^{tr}, B, in_B) \\
\mathbf{E}_3' &= (D_2^{tr}, A, in_A) \\
\mathbf{E}_4' &= (D_2^{tr}, C, (B, A), h, (\mathcal{I}_1', \mathcal{I}_1'))
\end{aligned}$$

where $D_2^{tr} = \{-1, \ldots, n - 5\}$, $in_B(i) = in_G(i + 5)$, $in_A(i) = in_X(i + 5)$, $h(a, b) = a - (b \bmod 4)$, and $\mathcal{I}_1'(i) = \mathcal{I}_1(i + 5) - 5 = i$.

   The data dependence graph and reduced dependence graphs of the system are illustrated in Fig. 4.6 c) and d), respectively.   ∎ 4.4.3

4.4.2 Regularisation Directions

Similar to the integral case, given a finitely generated data dependence, we will choose its regularisation directions among the generators of its index mapping. Because of the assumptions made in Section 4.3.3, that the coefficients of the index mapping are non-negative variables and the cone defined by the generators of the dependence mapping is pointed, this choice will guarantee the preservation of affine scheduling. We will address this issue in Section 4.5.

4.4.3 Injectivity of an Atomic Finitely Generated Index Mapping

We can avoid data broadcasts by restricting ourselves to injective index mapping (see Section 2.3.2). Given an atomic finitely generated index mapping $\mathcal{I}$ and a domain D in $\mathbf{Z}^n$, the injectivity of the index mappings $\mathcal{I}[in]$, for all inputs in, can be established by considering the geometric relation between D and the generator of $\mathcal{I}$. Indeed, this fact generalises the similar result for atomic integral index mappings (given in Proposition 3.2.4).

Proposition 4.4.4   Consider an atomic finitely generated index mapping $\mathcal{I}$ and a domain $D \subseteq \mathbf{Z}^n$, such that for all $z \in D$, $\mathcal{I}(z) = z + G(z)d$. If $d \notin lin(D)$ then for all $in \in Input$, $\mathcal{I}[in]$ is injective over D.

PROOF: For $in \in Input$, $\mathcal{I}[in]$ is the integral index mapping defined by $\mathcal{I}[in](z) = z + G[in](z)d$. From linear algebra, if $z, z' \in D$ then $z - z' \in lin(D)$. Assume that there exist $z, z' \in D$, such that $z \neq z'$ and $\mathcal{I}[in](z) = \mathcal{I}[in](z')$. We want to prove that this assumption always implies a contradiction with respect to the hypotheses of the proposition and therefore for all $z, z' \in D$, $z \neq z'$ implies $\mathcal{I}[in](z) \neq \mathcal{I}[in](z')$. There are only two possibilities, both leading to a contradiction. If $G[in](z) = G[in](z') = c$, then $\mathcal{I}[in](z) = \mathcal{I}[in](z')$ implies $z + cd = z' + cd$, i.e., $z = z'$. Otherwise, if $G[in](z) \neq G[in](z')$, then $G[in](z') - G[in](z) = c \neq 0$ and $\mathcal{I}[in](z) = \mathcal{I}[in](z')$ implies $z - z' = cd$, i.e., $d \in lin(D)$.
∎ 4.4.4

Example 4.4.5   Consider the data dependence $\mathcal{DD} = (D, U, V, \mathcal{I})$, with index mapping $\mathcal{I}(i, j) = (i, j) + G(i, j)(1, 1)$, domain $D = \{(i, j) \mid 1 \leq i \leq n, j = 2\}$, for some $n \in \mathbf{N}$, and G defined by the equation $\mathbf{E} = (D, G, in_G)$, with $in_G(i, j) \in \mathbf{Z}$ for all $(i, j) \in D$.

The generator of $\mathcal{I}$ is $d = (1, 1)$, $lin(D) = \langle (1, 0) \rangle$ and $d \notin lin(D)$ (see Fig. 4.7 a)). Hence, according to Proposition 4.4.4, all index mappings $\mathcal{I}[in]$,

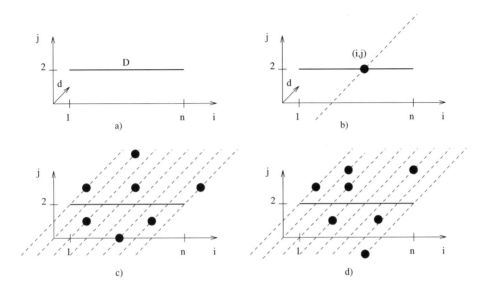

Fig. 4.7. Injectivity of an atomic finitely generated index mapping.

for all inputs, are injective on D. In fact, for all inputs in, $(i, j) + G[in](i, j)d$ belongs to the line $\{(i, j) + ld \mid l \in \mathbf{R}\}$ (see Fig. 4.7 b)). In addition, such lines are parallel for all points in D. This is illustrated in Fig. 4.7 c) and d) for two possible inputs of G.   ■ 4.4.5

4.4.4   Uniformisation

Uniformisation substitutes an atomic finitely generated data dependence with a system of uniform data dependencies. Let $\mathcal{DD} = (D, U, V, \mathcal{I})$ be an atomic finitely generated data dependence, with $\mathcal{I}(z) = z + G(z)d$. The application of uniformisation techniques is subject to the following constraints:

- that the coefficient G of the index mapping $\mathcal{I}$ is bounded over D. This restriction also applies to the integral case and its justification was discussed in Section 3.2.3;

- that variables G and U are separable. This property allows us to translate in the lattice space, the system of equations defining the variable G, so that the computations of G can be performed prior to those of U. The property also guarantees that no cycles are generated in the data dependence graph for effect of the translation.

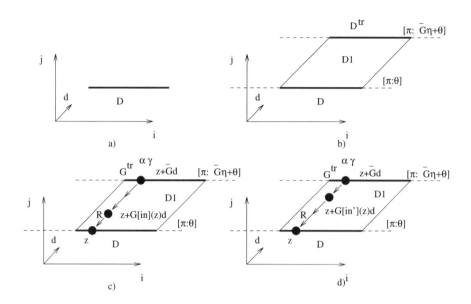

Fig. 4.8. Uniformisation 1.

Note that we are implicitly assuming that the index mapping has been specified correctly, hence that G is defined at each point of D. If this is not the case, the assumption we made in Section 4.1 on the default value of an integer-valued indexed variable implies that our techniques are still well-defined. This fact should be taken into account if verification issues are considered. To this end, a different choice of default value, which allows to identify a condition of error, may be more appropriate.

As for integral uniformisation, in the finitely generated case two uniformisation techniques are defined based on the geometric relation between the direction vector d and the domain D of the data dependence.

The first technique is illustrated in Fig. 4.8 for a 2-dimensional case. In the figure, $\bar{G}$ denotes the upper bound of G on D. Routing directions and domains are chosen similar to the integral case, and two control variables, α and γ, are required. In the figure, the vector π defines the hyperplane $[\pi : \theta]$ containing D and $\eta = \pi \cdot d$. For each z in D, α and γ are initialised at the point $z + \bar{G}d$. In contrast to the integral case, their initialisation requires the redefinition of the computations of G. For simplicity, let us assume that D is the definition domain of G. Then its computations are redefined on the translated domain D^{tr}, shown in Fig. 4.8 b). A new variable G^{tr} is used such that for each z in D, $G^{tr}(z + \bar{G}d)$ is defined to be the same as $G(z)$. The

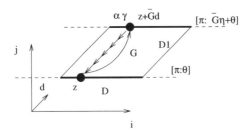

Fig. 4.9. The need for the redefinition of the coefficient G.

initialisation of α and γ is obtained through variable G^{tr} instead of G. The routing scheme is developed as for integral data dependencies. Fig. 4.8 c) and d) illustrate the routing scheme for two possible values of $G(z)$. That the scheme works for all inputs follows from the assumption that G is a non-negative bounded variable on D. In fact, for all inputs in and for all $z \in D$, $z + G[in](z)d$ lies on the segment $\{z + ld \mid 0 \leq l \leq \bar{G}\}$. On this segment the dynamic reconfiguration of the control variables α and γ guarantees that, for all inputs, the right data is transferred. The redefinition of variable G is necessary to avoid cycles in the resulting data dependence graph. If G were not redefined as G^{tr}, the resulting data dependence graph would resemble the cyclic graph sketched in Fig. 4.9.

Proposition 4.4.6 *[Uniformisation 1]* Let $\mathcal{DD} = (D, U, V, \mathcal{I})$ be an atomic finitely generated data dependence, with $\mathcal{I}(z) = z + G(z)d$. Let G be non-negative and bounded over D and let $\bar{G}$ be the least upper bound of G. Let $d \notin lin(D)$ and P be the convex polyhedron generated by D and d. Consider $\pi \in lin(P) \cap D^{\perp}$, with $\pi \neq 0$, and the hyperplane $[\pi : \theta]$ containing the domain D. Let U and G be separable. Consider the system of equations $(Def\mathbf{S}_G)^{tr,ren}$, defined by the renaming $ren(G) = G^{tr}$ and the translation $tr(z) = z + \bar{G}d$.

Then $\mathcal{DD}$ can be substituted by the uniform data dependence

$$\mathcal{DD}' = (D, U, R, \mathcal{I}_0)$$

and the system of equations:

$$
\begin{aligned}
\mathbf{E}_1 &= (D_1, R, (R, V, \alpha, \gamma), f, (\mathcal{I}_1, \mathcal{I}_0, \mathcal{I}_0, \mathcal{I}_0)) \\
\mathbf{E}_2 &= (D_{1,1}, \alpha, \alpha, id, \mathcal{I}_1) \\
\mathbf{E}_3 &= (D_{1,2}, \alpha, G^{tr}, id, \mathcal{I}_0)
\end{aligned}
$$

$$\mathbf{E}_4 = (D_{1,1}, \gamma, \gamma, dec, \mathcal{I}_1)$$
$$\mathbf{E}_5 = (D_{1,2}, \gamma, in_\gamma)$$

where:

- the index mappings are:

$$\mathcal{I}_0(z) = z$$
$$\mathcal{I}_1(z) = z + d$$

- R, α and γ are new variables;
- the applied functions are:

$$in_\gamma(z) = \bar{G}$$
$$id(a) = a$$
$$dec(a) = a - 1$$
$$f(a, b, c, d) = \begin{cases} a & c \neq d \\ b & \text{otherwise} \end{cases}$$

- the new domains are:

$$D_1 = \{z + ld \mid z \in D, 0 \leq l \leq \bar{G}\}$$
$$D_{1,1} = \{z + ld \mid z \in D, 0 \leq l < \bar{G}\}$$
$$D_{1,2} = \{z + \bar{G}d \mid z \in D\}$$

PROOF: The proof relies upon the fact that, for all $z \in D$ and for all inputs in, $G[in](z) \in \{0, \ldots, \bar{G}\}$.

For all $z \in D$, let $segm(z) = \{z + ld \mid 0 \leq l \leq \bar{G}\}$. By definition, for all $z' \in segm(z)$, $\alpha(z') = G(z)$. In fact, for $z \in D$,

$$\alpha(z) = \alpha(z + d)$$
$$= \ldots = \alpha(z + \bar{G}d) = G^{tr}(z + \bar{G}d) = G(z).$$

Also for all $z' = z + ld \in segm(z)$, $\gamma(z') = l$. In fact, for $z \in D$,

$$\gamma(z) = \gamma(z + d) - 1$$
$$= \ldots = \gamma(z + \bar{G}d) - \bar{G} = \bar{G} - \bar{G} = 0.$$

Therefore, for all $z' = z + ld \in segm(z)$, $\alpha(z') = \gamma(z')$ if and only if $l = G(z)$.

Hence, for all $z \in D$,

$$U(z) = R(\mathcal{I}_0(z)) = R(\mathcal{I}_1 \circ \mathcal{I}_0(z))$$
$$= \quad \ldots = R(\mathcal{I}_1^{G(z)} \circ \mathcal{I}_0(z)) = V(\mathcal{I}_0 \circ \mathcal{I}_1^{G(z)} \circ \mathcal{I}_0(z))$$
$$= \quad V(z + G(z)d) = V(\mathcal{I}(z)).$$

■ 4.4.6

Corollary 4.4.7   Let $\mathcal{DD} = (D, U, V, \mathcal{I})$ be an atomic finitely generated data dependence, with $\mathcal{I}(z) = z + G(z)d$. Let G be non-negative and bounded over D and $\bar{G}$ the least upper bound of G.

If $\bar{G} = 0$, then $\mathcal{DD}$ can be substituted by the following uniform data dependence

$$\mathcal{DD}_0 \quad = \quad (D, U, V, \mathcal{I}_0)$$

where $\mathcal{I}_0(z) = z$.   ■ 4.4.7

The second uniformisation technique consists of the combination of the corresponding integral techniques with the redefinition of the computations of the coefficient G of the index mapping. Once again, for simplicity, let us assume that D is the definition domain of G. The technique is illustrated in Fig. 4.10 for a 3-dimensional case. In the figure D^{tr} indicates the translated domain and G^{tr} the new variable defined as G on D^{tr}. Fig. 4.10 c) and d) illustrate the routing scheme for two possible values of $G(z)$. Regularisation directions and domains, as well as routing and control variables are defined as for integral uniformisation.

Proposition 4.4.8 *[Uniformisation 2]* Let $\mathcal{DD} = (D, U, V, \mathcal{I})$ be an atomic finitely generated data dependence, with $\mathcal{I}(z) = z + G(z)d$. Let G be non-negative and bounded over D, m the least upper bound of G and $\bar{G} = \lfloor m/2 \rfloor$. Let $d \in lin(D)$ and $dim(D) < n$. Consider $\pi \in D^{\perp}$, with $\pi \neq 0$, the hyperplane $[\pi : 0]$ containing the domain D, and let $\hat{d} = d + \pi$ and $\check{d} = d - \pi$. Let U and G be separable. Consider the system of equations $(Def\mathbf{S}_G)^{tr,ren}$, defined by the renaming $ren(G) = G^{tr}$ and the translation $tr(z) = z + \bar{G}\hat{d}$.

Then $\mathcal{DD}$ can be substituted by the uniform data dependence

$$\mathcal{DD}' \quad = \quad (D, U, R^1, \mathcal{I}_0)$$

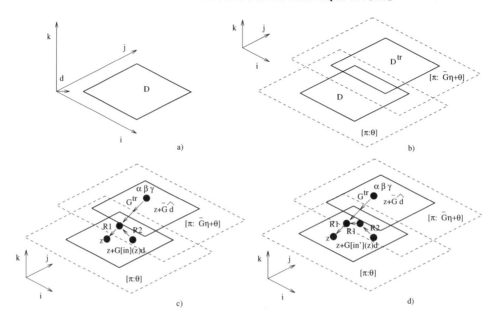

Fig. 4.10. Uniformisation 2.

and the system of equations:

$$
\begin{aligned}
\mathbf{E}_1 &= (D_1, R^1, (R^1, R^2, R^2, \alpha, \beta, \gamma), f, (\mathcal{I}_1, \mathcal{I}_{2,0}, \mathcal{I}_{2,1}, \mathcal{I}_0, \mathcal{I}_0, \mathcal{I}_0)) \\
\mathbf{E}_2 &= (D_{2,1}, R^2, R^2, id, \mathcal{I}_3) \\
\mathbf{E}_3 &= (D_{2,2}, R^2, V, id, \mathcal{I}_0) \\
\mathbf{E}_4 &= (D_{1,1}, \alpha, \alpha, id, \mathcal{I}_1) \\
\mathbf{E}_5 &= (D_{1,2}, \alpha, G^{tr}, half\_floor, \mathcal{I}_0) \\
\mathbf{E}_6 &= (D_{1,1}, \beta, \beta, id, \mathcal{I}_1) \\
\mathbf{E}_7 &= (D_{1,2}, \beta, G^{tr}, mod_2, \mathcal{I}_0) \\
\mathbf{E}_8 &= (D_{1,1}, \gamma, \gamma, dec, \mathcal{I}_1) \\
\mathbf{E}_9 &= (D_{1,2}, \gamma, in_\gamma)
\end{aligned}
$$

where:

- the index mappings are:

$$
\begin{aligned}
\mathcal{I}_0(z) &= z \\
\mathcal{I}_1(z) &= z + \hat{d} \\
\mathcal{I}_{2,0}(z) &= z
\end{aligned}
$$

$$\begin{aligned}
\mathcal{I}_{2,1}(z) &= z + d \\
\mathcal{I}_3(z) &= z + \check{d}
\end{aligned}$$

- R^1, R^2, α, β and γ are new variables;

- the applied functions are:

$$\begin{aligned}
in_\gamma(z) &= \bar{G} \\
id(a) &= a \\
half\_floor(a) &= \lfloor a/2 \rfloor \\
mod_2(a) &= a \bmod 2 \\
dec(a) &= a - 1 \\
f(a,b,c,d,e,f) &= \begin{cases} a & d \neq f \\ b & d = f, e = 0 \\ c & d = f, e = 1 \end{cases}
\end{aligned}$$

- the new domains are:

$$\begin{aligned}
D_1 &= \{z + l\hat{d} \mid z \in D, 0 \leq l \leq \bar{G}\} \\
D_{1,1} &= \{z + l\hat{d} \mid z \in D, 0 \leq l < \bar{G}\} \\
D_{1,2} &= \{z + \bar{G}\hat{d} \mid z \in D\} \\
D_2 &= \{z + l_1 d + l_2 \check{d} \mid z \in D_1, 0 \leq l_1 \leq 1, 0 \leq l_2 \leq \bar{G}\} \cap \\
&\quad \{z \in \mathbf{Z}^n \mid \pi \cdot z \geq \theta\} \\
D_{2,1} &= \{z \in D_2 \mid \pi \cdot z > \theta\} \\
D_{2,2} &= \{z \in D_2 \mid \pi \cdot z = \theta\}.
\end{aligned}$$

PROOF: The proof relies upon the fact that, for all $z \in D$ and for all inputs in, $G[in](z) \in \{0, \ldots, \bar{G}\}$.

For all $z \in D$, let $segm(z) = \{z + ld \mid 0 \leq l \leq \bar{G}\}$. By definition, for all $z' \in segm(z)$, $\alpha(z') = \lfloor G(z)/2 \rfloor$. In fact, for $z \in D$,

$$\begin{aligned}
\alpha(z) = \alpha(z + \hat{d}) = \ldots &= \alpha(z + \bar{G}\hat{d}) \\
&= half\_floor(G^{tr}(z + \bar{G}\hat{d})) = half\_floor(G(z)) = \lfloor G(z)/2 \rfloor.
\end{aligned}$$

Also, for all $z' \in segm(z)$, $\beta(z') = G(z) \bmod 2$. In fact, for $z \in D$,

$$\begin{aligned}
\beta(z) = \beta(z + \hat{d}) = \ldots &= \beta(z + \bar{G}\hat{d}) \\
&= mod_2(G^{tr}(z + \bar{G}\hat{d})) = mod_2(G(z)) = G(z) \bmod 2.
\end{aligned}$$

Finally, for all $z' = z + l\hat{d} \in segm(z)$, $\gamma(z') = l$. In fact, for $z \in D$,

$$\gamma(z) = \gamma(z + \hat{d}) - 1$$
$$= \quad \ldots = \gamma(z + \bar{G}\hat{d}) - \bar{G} = \bar{G} - \bar{G} = 0.$$

Therefore, for all $z' = z + ld \in segm(z)$, $\alpha(z') = \gamma(z')$ if and only if $l = \lfloor G(z)/2 \rfloor$.

We observe that, for all $c \in \mathbf{Z}$,

$$\lfloor c/2 \rfloor = (c - c \bmod 2)/2$$
$$c = 2\lfloor c/2 \rfloor + c \bmod 2.$$

Hence, for all $z \in D$,

$$
\begin{aligned}
U(z) &= R^1(\mathcal{I}_0(z)) = R^1(\mathcal{I}_1 \circ \mathcal{I}_0(z)) \\
&= \ldots = R^1(\mathcal{I}_1^{\lfloor G(z)/2 \rfloor} \circ \mathcal{I}_0(z)) \\
&= R^2(\mathcal{I}_{2,G(z) \bmod 2} \circ \mathcal{I}_1^{\lfloor G(z)/2 \rfloor} \circ \mathcal{I}_0(z)) \\
&= R^2(\mathcal{I}_3 \circ \mathcal{I}_{2,G(z) \bmod 2} \circ \mathcal{I}_1^{\lfloor G(z)/2 \rfloor} \circ \mathcal{I}_0(z)) \\
&= \ldots = R^2(\mathcal{I}_3^{\lfloor G(z)/2 \rfloor} \circ \mathcal{I}_{2,G(z) \bmod 2} \circ \mathcal{I}_1^{\lfloor G(z)/2 \rfloor} \circ \mathcal{I}_0(z)) \\
&= V(\mathcal{I}_0 \circ \mathcal{I}_3^{\lfloor G(z)/2 \rfloor} \circ \mathcal{I}_{2,G(z) \bmod 2} \circ \mathcal{I}_1^{\lfloor G(z)/2 \rfloor} \circ \mathcal{I}_0(z)) \\
&= V(z + \lfloor G(z)/2 \rfloor \hat{d} + (G(z) \bmod 2)d + \lfloor G(z)/2 \rfloor \check{d}) \\
&= V(z + 2\lfloor G(z)/2 \rfloor d + (G(z) \bmod 2)d) \\
&= V(z + (2\lfloor G(z)/2 \rfloor + G(z) \bmod 2)d) \\
&= V(z + G(z)d) = V(\mathcal{I}(z)).
\end{aligned}
$$

■ 4.4.8

Corollary 4.4.9   Let $\mathcal{DD} = (D, U, V, \mathcal{I})$ be an atomic finitely generated data dependence, with $\mathcal{I}(z) = z + G(z)d$. Let G be non-negative and bounded over D, m the least upper bound of G, and $\bar{G} = \lfloor m/2 \rfloor$.

If $\bar{G} = 0$ then $\mathcal{DD}$ can be substituted by the uniform data dependence:

$$\mathcal{DD}' = (D, U, R, \mathcal{I}_0)$$

and the equations:

$$
\begin{aligned}
\mathbf{E}_1 &= (D, R, (V, V, \beta), f, (\mathcal{I}_0, \mathcal{I}_1, \mathcal{I}_0)) \\
\mathbf{E}_2 &= (D, \beta, G, mod_2, \mathcal{I}_0)
\end{aligned}
$$

where:

- the index mappings are:

$$\begin{aligned}
\mathcal{I}_0(z) &= z \\
\mathcal{I}_1(z) &= z + d
\end{aligned}$$

- R and β are new variables;

- the applied functions are:

$$\begin{aligned}
mod_2(a) &= a \bmod 2 \\
f(a, b, c) &= \begin{cases} a & c = 0 \\ b & c = 1 \end{cases}
\end{aligned}$$

■ 4.4.9

Example 4.4.10   Consider the system **S** of equations of Example 4.2.2. As Y and I are separable, we can apply uniformisation to the data dependence $\mathcal{DD} = (D_2, Y, Y, \mathcal{I}_2)$, with $\mathcal{I}_2(i) = i - I(i)$ and $D_2 = \{4, \ldots, n\}$.

As $lin(D) = \mathbf{Z}$, $d = (-1) \in lin(D)$ and $dim(D) = 1$ we need to reindex (see Section 2.2.2) the equations in $\mathbf{Z}^2$ (so that $dim(D) < 2$). Let $\pi = (0, 1)$, $\hat{d} = d + \pi = (-1, 1)$ and $\check{d} = d - \pi = (-1, -1)$. Note that $I(i) \leq n$ for all $i \in \{4, \ldots, n\}$. Consider the translation $tr(i, j) = (i, j) + \lfloor n/2 \rfloor \hat{d} = (i - \lfloor n/2 \rfloor, j + \lfloor n/2 \rfloor)$, and the variable renaming ren such that $ren(G) = G^{tr}$, $ren(X) = X^{tr}$ and $ren(I) = I^{tr}$.

The application of Proposition 4.4.8 and the generation of the corresponding equations produces the following system:

$$\begin{aligned}
&\qquad\qquad (\text{inputs of } Y) \\
\mathbf{E}_1 &= (D_0, Y, in_Y) \\
&\qquad\qquad (\text{translated sub-system of equations}) \\
\mathbf{E}_2 &= (D^{tr}, G^{tr}, in_G) \\
\mathbf{E}_3 &= (D^{tr}, X^{tr}, in_X) \\
\mathbf{E}_4 &= (D^{tr}, I^{tr}, (X^{tr}, G^{tr}), h, (\mathcal{I}_1, \mathcal{I}_1)) \\
&\qquad\qquad (\text{computations of } Y) \\
\mathbf{E}_5 &= (D, Y, R^1, sqr, \mathcal{I}_1) \\
&\qquad\qquad (\text{routing variables } R^1, R^2) \\
\mathbf{E}_6 &= (D_1, R^1, (R^1, R^2, R^2, \alpha, \beta, \gamma), f, (\mathcal{I}_2, \mathcal{I}_1, \mathcal{I}_3, \mathcal{I}_1, \mathcal{I}_1, \mathcal{I}_1))
\end{aligned}$$

$$\begin{aligned}
\mathbf{E}_7 &= (D_{2,1}, R^2, R^2, id, \mathcal{I}_4) \\
\mathbf{E}_8 &= (D_{2,2}, R^2, Y, id, \mathcal{I}_1) \\
&\quad \text{(control signals } \alpha, \beta, \gamma) \\
\mathbf{E}_9 &= (D_{1,1}, \alpha, \alpha, id, \mathcal{I}_2) \\
\mathbf{E}_{10} &= (D_{1,2}, \alpha, I^{tr}, half\_floor, \mathcal{I}_1) \\
\mathbf{E}_{11} &= (D_{1,1}, \beta, \beta, id, \mathcal{I}_2) \\
\mathbf{E}_{12} &= (D_{1,2}, \beta, I^{tr}, mod_2, \mathcal{I}_1) \\
\mathbf{E}_{13} &= (D_{1,1}, \gamma, \gamma, dec, \mathcal{I}_2) \\
\mathbf{E}_{14} &= (D_{1,2}, \gamma, in_\gamma)
\end{aligned}$$

where:

- the index mappings are:

$$\begin{aligned}
\mathcal{I}_1(i,j) &= (i,j) \\
\mathcal{I}_2(i,j) &= (i,j) + \hat{d} = (i-1, j+1) \\
\mathcal{I}_3(i,j) &= (i,j) + d = (i-1, j) \\
\mathcal{I}_4(i,j) &= (i,j) + \check{d} = (i-1, j-1)
\end{aligned}$$

- the applied functions are $half\_floor(a) = \lfloor a/2 \rfloor$, $mod_2(a) = a \bmod 2$, $id(a) = a$, $dec(a) = a - 1$ $in_\gamma(i,j) = \lfloor n/2 \rfloor$, $h(a,b) = a - (b \bmod 4)$, and

$$f(a, b, c, \alpha, \beta, \gamma) = \begin{cases} a & \alpha \neq \gamma \\ b & \alpha = \gamma, \beta = 0 \\ c & \alpha = \gamma, \beta = 1 \end{cases}$$

- the domains are:

$$\begin{aligned}
D_0 &= \{(i,j) \mid 0 \leq i \leq 3, j = 0\} \\
D &= \{(i,j) \mid 4 \leq i \leq n, j = 0\} \\
D^{tr} &= \{(i,j) \mid 4 - \lfloor n/2 \rfloor \leq i \leq n - \lfloor n/2 \rfloor, j = \lfloor n/2 \rfloor\} \\
D_1 &= \{(i,j) \mid 4 - j \leq i \leq n - j, 0 \leq j \leq \lfloor n/2 \rfloor\} \\
D_{1,1} &= \{(i,j) \mid 4 - j \leq i \leq n - j, 0 \leq j \leq \lfloor n/2 \rfloor - 1\} \\
D_{1,2} &= \{(i,j) \mid 4 - \lfloor n/2 \rfloor \leq i \leq n - \lfloor n/2 \rfloor, j = \lfloor n/2 \rfloor\} \\
D_2 &= \{(i,j) \mid 2 - 2\lfloor n/2 \rfloor + j \leq i \leq n - j, 0 \leq j \leq \lfloor n/2 \rfloor\} \\
D_{2,1} &= \{(i,j) \in D_2 \mid j > 0\} \\
D_{2,2} &= \{(i,j) \in D_2 \mid j = 0\}
\end{aligned}$$

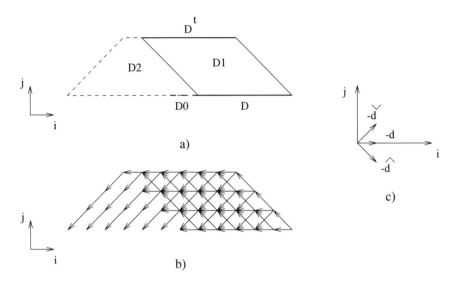

Fig. 4.11. a) Domains after uniformisation; b) Data dependence graph; c) routing direction vectors.

The resulting domains, data dependence graph and routing direction vectors are illustrated in Fig. 4.11 a), b) and c), respectively. A possible corresponding code segment is the following (that the reader may compare with that initially given in Example 4.2.2):

$n' := $ `floor`$(n/2)$;
(* separated code segment *)
`for` $i := 4 - n'$ `to` $n - n'$ `do`
  `begin`
    `read`$(G^{tr}(i))$;
    $X^{tr}(i) := i + n'$;
  $I^{tr}(i) := X^{tr}(i) - G^{tr}(i)$ `mod` 4;
    `end`;
(* control signals *)
`for` $i := 4 - n'$ `to` $n - n'$ `do`
  `begin`
    $\alpha(i, n') := $ `floor`$(I^{tr}(i)/2)$;
    $\beta(i, n') := I^{tr}(i)$ `mod` 2;
    $\gamma(i, n') := n'$;
    `for` $j := 1$ `to` n `do`

```
        begin
            α(i + j, n' − j) := α(i + (j − 1), n' − (j − 1));
            β(i + j, n' − j) := β(i + (j − 1), n' − (j − 1));
            γ(i + j, n' − j) := γ(i + (j − 1), n' − (j − 1)) − 1;
        end;
    end;
    (* inputs of Y and their routing *)
    for i := 0 to 3 do
        begin
            read(Y(i));
            R²(i, 0) := Y(i);
            for j := 1 to floor(I(i)/2) do
                R²(i + j, j) := R²(i + j − 1), j − 1);
        end;
    (* computations of Y and their routing *)
    for i := 4 to n do
        begin
            R¹(i − n' − 1, n' + 1) :=⊥;
            for j := 0 to n' do
                if (α(i − n' + j, n' − j) ≠ γ(i − n' + j, n' − j))
                then R¹(i − n' + j, n' − j) := R¹(i − n' + j − 1, n' − j + 1);
                else if (β(i − n' + j, n' − j) = 1)
                    then R¹(i − n' + j, n' − j) := R²(i − n' + j − 1, n' − j);
                    else R¹(i − n' + j, n' − j) := R²(i − n' + j, n' − j);
            Y(i) := R¹(i, 0);
            R²(i, 0) := Y(i);
            for j := 1 to n' do
                R²(i + j, j) := R²(i + j − 1, j − 1);
        end;                                        ■ 4.4.10
```

4.4.5   Parametric Uniformisation

Parametric uniformisation techniques can be used to reduce routing overhead by allowing a restricted amount of overloading of the data dependence
graph (see Sections 2.3.2 and 3.2.4 for further discussion). A technique
similar to that for atomic integral data dependencies can be developed for
atomic finitely generated data dependencies, the main difference being the
separation of the computations of the coefficient of the index mapping. The

reader is referred to the discussion in Section 3.2.4 for a description of the technique.

Proposition 4.4.11 [*Parametric Uniformisation*] Let $\mathcal{DD} = (D, U, V, \mathcal{I})$ be an atomic finitely generated data dependence, with $\mathcal{I}(z) = z + G(z)d$. Let G be non-negative and bounded over D and m the least upper bound of G. Let $p \in \mathbf{N}^+$ and $\bar{G} = \lfloor m/(p+1) \rfloor$. Let $d \in lin(D)$ and $dim(D) < n$. Consider $\pi \in D^\perp$, with $\pi \neq 0$, the hyperplane $[\pi : 0]$ containing the domain D, and let $\hat{d} = d + \pi$ and $\check{d} = d - \pi$. Let U and G be separable. Consider the system of equations $(Def\mathbf{S}_G)^{tr,ren}$, defined by the renaming $ren(G) = G^{tr}$ and the translation $tr(z) = z + \bar{G}\hat{d}$.

Then $\mathcal{DD}$ can be substituted by the uniform data dependence

$$\mathcal{DD}' \ = \ (D, U, R^1, \mathcal{I}_0)$$

and the system of equations:

$$
\begin{aligned}
\mathbf{E}_1 &= (D_1, R^1, (R^1, R^2, \ldots, R^2, \alpha, \beta, \gamma), f, (\mathcal{I}_1, \mathcal{I}_{2,0}, \ldots, \mathcal{I}_{2,p}, \mathcal{I}_0, \mathcal{I}_0, \mathcal{I}_0)) \\
\mathbf{E}_2 &= (D_{2,1}, R^2, R^3, id, \mathcal{I}_3) \\
\mathbf{E}_3 &= (D_{2,2}, R^2, V, id, \mathcal{I}_0) \\
\mathbf{E}_4 &= (D_{2,3}, R^3, R^2, id, \mathcal{I}_4) \\
\mathbf{E}_5 &= (D_{1,1}, \alpha, \alpha, id, \mathcal{I}_1) \\
\mathbf{E}_6 &= (D_{1,2}, \alpha, G^{tr}, (p+1)\_floor, \mathcal{I}_0) \\
\mathbf{E}_7 &= (D_{1,1}, \beta, \beta, id, \mathcal{I}_1) \\
\mathbf{E}_8 &= (D_{1,2}, \beta, G^{tr}, mod_{p+1}, \mathcal{I}_0) \\
\mathbf{E}_9 &= (D_{1,1}, \gamma, \gamma, dec, \mathcal{I}_1) \\
\mathbf{E}_{10} &= (D_{1,2}, \gamma, in_\gamma)
\end{aligned}
$$

where:

- the index mappings are:

$$
\begin{aligned}
\mathcal{I}_0(z) &= z \\
\mathcal{I}_1(z) &= z + \hat{d} \\
\mathcal{I}_{2,0}(z) &= z \\
\mathcal{I}_{2,1}(z) &= z + d \\
&\cdots \\
\mathcal{I}_{2,p}(z) &= z + pd
\end{aligned}
$$

$$\begin{aligned}
\mathcal{I}_3(z) &= z + \check{d} \\
\mathcal{I}_4(z) &= z + (p-1)d
\end{aligned}$$

- $R^1, R^2, R^3, \alpha, \beta$ and γ are new variables;

- the applied functions are:

$$\begin{aligned}
in_\gamma(z) &= \bar{G} \\
id(a) &= a \\
(p+1)\_floor(a) &= \lfloor a/(p+1) \rfloor \\
mod_{p+1}(a) &= a \bmod (p+1) \\
dec(a) &= a - 1 \\
f(a, b_0, \ldots, b_p, c, d, e) &= \begin{cases}
a & c \neq e \\
b_0 & c = e, d = 0 \\
\ldots & \\
b_p & c = e, d = p
\end{cases}
\end{aligned}$$

- the new domains are:

$$\begin{aligned}
D_1 &= \{z + l\hat{d} \mid z \in D, 0 \leq l \leq \bar{G}\} \\
D_{1,1} &= \{z + l\hat{d} \mid z \in D, 0 \leq l < \bar{G}\} \\
D_{1,2} &= \{z + \bar{G}\hat{d} \mid z \in D\} \\
D_2 &= \{z + (l_1 + l_2)d + l_3\check{d} \mid z \in D_1, 0 \leq l_1 \leq p, 0 \leq l_2 \leq (p-1)\bar{G}, \\
&\qquad 0 \leq l_3 \leq \bar{G}\} \cap \{z \in \mathbf{Z}^n \mid \pi \cdot z \geq \theta\} \\
D_{2,1} &= \{z \in D_2 \mid \pi \cdot z > \theta\} \\
D_{2,2} &= \{z \in D_2 \mid \pi \cdot z = \theta\} \\
D_{2,3} &= \{z \in D_2 \mid \pi \cdot z < \theta + \bar{G}\eta\}.
\end{aligned}$$

PROOF: The proof relies upon the fact that for all $z \in D$ and for all inputs in, $G[in](z) \in \{0, \ldots, \bar{G}\}$.

For all $z \in D$, let $segm(z) = \{z + ld \mid 0 \leq l \leq \bar{G}\}$. By definition, for all $z' \in segm(z)$, $\alpha(z') = \lfloor G(z)/(p+1) \rfloor$. In fact, for $z \in D$,

$$\begin{aligned}
\alpha(z) = \alpha(z + \hat{d}) &= \ldots = \alpha(z + \bar{G}\hat{d}) \\
&= (p+1)\_floor(G^{tr}(z + \bar{G}\hat{d})) = (p+1)\_floor(G(z)) \\
&= \lfloor G(z)/(p+1) \rfloor.
\end{aligned}$$

Also, for all $z' = z + l\hat{d}$ in the segment $segm(z)$, with $z \in D$, $\beta(z') = G(z) \bmod (p+1)$. In fact, for $z \in D$,

$$\beta(z) = \beta(z + \hat{d}) = \ldots = \beta(z + \bar{G}\hat{d})$$
$$= mod_{p+1}(G^{tr}(z + \bar{G}\hat{d})) = mod_{p+1}(G(z)) = G(z) \bmod (p+1).$$

Finally, for all $z' = z + l\hat{d} \in segm(z)$, $\gamma(z') = l$. In fact, for $z \in D$,

$$\gamma(z) = \gamma(z + \hat{d}) - 1$$
$$= \ldots = \gamma(z + \bar{G}\hat{d}) - \bar{G} = \bar{G} - \bar{G} = 0.$$

Therefore, for all $z' = z + ld \in segm(z)$, $\alpha(z') = \gamma(z')$ if and only if $l = \lfloor G(z)/(p+1) \rfloor$.

We observe that, for all $c \in \mathbf{Z}$,

$$\lfloor c/(p+1) \rfloor = (c - c \bmod (p+1))/(p+1)$$
$$c = (p+1)\lfloor c/(p+1) \rfloor + c \bmod (p+1).$$

Hence, for all $z \in D$,

$$U(z) = R^1(\mathcal{I}_0(z)) = R^1(\mathcal{I}_1 \circ \mathcal{I}_0(z))$$
$$= \ldots = R^1(\mathcal{I}_1^{\lfloor G(z)/(p+1) \rfloor} \circ \mathcal{I}_0(z))$$
$$= R^2(\mathcal{I}_{2,(G(z) \bmod (p+1))} \circ \mathcal{I}_1^{\lfloor G(z)/(p+1) \rfloor} \circ \mathcal{I}_0(z))$$
$$= R^3(\mathcal{I}_3 \circ \mathcal{I}_{2,(G(z) \bmod (p+1))} \circ \mathcal{I}_1^{\lfloor G(z)/(p+1) \rfloor} \circ \mathcal{I}_0(z))$$
$$= R^2(\mathcal{I}_4 \circ \mathcal{I}_3 \circ \mathcal{I}_{2,(G(z) \bmod (p+1))} \circ \mathcal{I}_1^{\lfloor G(z)/(p+1) \rfloor} \circ \mathcal{I}_0(z))$$
$$= \ldots$$
$$= R^2(\mathcal{I}_4^{\lfloor G(z)/(p+1) \rfloor} \circ \mathcal{I}_3^{\lfloor G(z)/(p+1) \rfloor} \circ \mathcal{I}_{2,(G(z) \bmod (p+1))}$$
$$\circ \mathcal{I}_1^{\lfloor G(z)/(p+1) \rfloor} \circ \mathcal{I}_0(z))$$
$$= V(\mathcal{I}_0 \circ \mathcal{I}_4^{\lfloor G(z)/(p+1) \rfloor} \circ \mathcal{I}_3^{\lfloor G(z)/(p+1) \rfloor} \circ \mathcal{I}_{2,(G(z) \bmod (p+1))}$$
$$\circ \mathcal{I}_1^{\lfloor G(z)/(p+1) \rfloor} \circ \mathcal{I}_0(z))$$
$$= V(z + \lfloor G(z)/(p+1) \rfloor \hat{d} + (G(z) \bmod (p+1))d +$$
$$\lfloor G(z)/(p+1) \rfloor \check{d} + \lfloor G(z)/(p+1) \rfloor (p-1)d)$$
$$= V(z + ((p+1)\lfloor G(z)/(p+1) \rfloor + G(z) \bmod (p+1))d)$$
$$= V(z + G(z)d) = V(\mathcal{I}(z)).$$

$$\blacksquare \; 4.4.11$$

Corollary 4.4.12   Let $\mathcal{DD} = (D, U, V, \mathcal{I})$ be an atomic finitely generated data dependence, with $\mathcal{I}(z) = z + G(z)d$. Let G be non-negative and bounded over D and m the least upper bound of G. Let $p \in \mathbf{N}^+$ and $\bar{G} = \lfloor m/(p+1) \rfloor$.
     If $\bar{G} = 0$ then $\mathcal{DD}$ can be substituted by the uniform data dependence

$$\mathcal{DD}' \;=\; (D, U, R, \mathcal{I}_0)$$

and the equations:

$$
\begin{aligned}
\mathbf{E}_1 &= (D, R, (V, \ldots, V, \beta), f, (\mathcal{I}_{1,0}, \ldots, \mathcal{I}_{1,p}, \mathcal{I}_0)) \\
\mathbf{E}_2 &= (D, \beta, G, mod_{p+1}, \mathcal{I}_0)
\end{aligned}
$$

where:

   - the index mappings are:

$$
\begin{aligned}
\mathcal{I}_0(z) &= z \\
\mathcal{I}_{1,0}(z) &= z \\
\mathcal{I}_{1,1}(z) &= z + d \\
&\cdots \\
\mathcal{I}_{1,p}(z) &= z + pd
\end{aligned}
$$

   - R and β are new variables;

   - the applied functions are:

$$
\begin{aligned}
mod_{p+1}(a) &= a \; mod \; (p+1) \\
f(a_0, \ldots, a_p, b) &= \begin{cases} a_0 & b = 0 \\ a_1 & b = 1 \\ \cdots & \\ a_p & b = p \end{cases}
\end{aligned}
$$

■ 4.4.12

4.4.6   Decomposition

Decomposition techniques allow us to substitute a finitely generated data dependence with a number of generators $m > 1$, with a corresponding finite set of atomic finitely generated data dependencies. Once again, the techniques derive from those for integral data dependencies by taking the separation of the coefficients of the index mappings into account. Note,

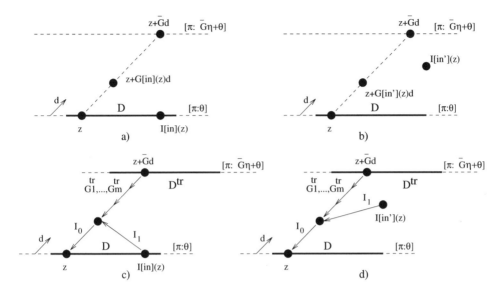

Fig. 4.12. Decomposition 1.

however, that because of the translation of their subsystems of equations, no inverse mappings are needed in the transformations. In particular, let $\mathcal{DD} = (D, U, V, \mathcal{I})$ be a finitely generated data dependence with index mapping $\mathcal{I}(z) = z + G(z)d + \sum_{j=1}^{m} G_j(z)d_j$. Two decomposition techniques are developed.

The first technique applies when $d \notin lin(D)$ and is illustrated in Fig. 4.12 (in 2 dimensions). For all z in D, $\mathcal{I}(z)$ and $z + G(z)d$ assume different values for different inputs. Two of these values are illustrated in parts a) and b) of the figure. However, for all inputs, $z + G(z)d$ lies on the segment $\{z + ld \mid 0 \leq l \leq \bar{G}\}$, where $\bar{G}$ is an upper bound of G on D. In Fig. 4.12 a) and b) the segment is represented as a dashed line through z. The expression $z + G(z)d$ defines the atomic finitely generated index mapping $\mathcal{I}_0$ of the decomposition. New variables G_j^{tr} are defined such that for all $z \in D$, $G_j^{tr}(z + \hat{G}d)$ assumes the same value as $G_j(z)$. Their values are pipelined from $z + \hat{G}d$ to z along the direction of $-d$, and used in the definition of the finitely generated index mapping $\mathcal{I}_1$ of the decomposition (see Fig. 4.12 c) and d)). Hence, an inverse of $\mathcal{I}_0$ is not necessary for the definition of $\mathcal{I}_1$ (it is instead for integral decomposition – see Proposition 3.2.19).

Proposition 4.4.13 *[Decomposition 1]* Let $\mathcal{DD} = (D, U, V, \mathcal{I})$ be a finitely generated data dependence, with $\mathcal{I}(z) = z + G(z)d + \sum_{j=1}^{m} G_j(z)d_j$. Let

G be non-negative and bounded over D, and $\bar{G}$ the least upper bound of G. Let $d \notin lin(D)$ and P be the convex polyhedron generated by D and d. Consider $\pi \in lin(P) \cap D^{\perp}$, with $\pi \neq 0$, and the hyperplane $[\pi : 0]$ containing the domain D. Let U and G_j be separable, for all j. Consider the system of equations $(Def\mathbf{S}_{\{G_1,...,G_m\}})^{tr,ren}$, defined by the renaming $ren(G_j) = G_j^{tr}$ and the translation $tr(z) = z + \bar{G}d$.

Then $\mathcal{DD}$ can be substituted by the atomic finitely generated data dependence

$$\mathcal{DD}' = (D, U, R, \mathcal{I}_0)$$

and the equations:

$$
\begin{aligned}
\mathbf{E}_1 &= (D_1, R, V, id, \mathcal{I}_1) \\
\mathbf{E}_{1_1} &= (D_{1,1}, G'_j, G'_1, id, \mathcal{I}_2) \\
\mathbf{E}_{1_2} &= (D_{1,2}, G'_j, G_1^{tr}, id, \mathcal{I}_3)
\end{aligned}
$$

$$\cdots$$

$$
\begin{aligned}
\mathbf{E}_{m_1} &= (D_{1,1}, G'_j, G'_m, id, \mathcal{I}_2) \\
\mathbf{E}_{m_2} &= (D_{1,2}, G'_j, G_m^{tr}, id, \mathcal{I}_3)
\end{aligned}
$$

where:

- the index mappings are:

$$
\begin{aligned}
\mathcal{I}_0(z) &= z + G(z)d \\
\mathcal{I}_1(z) &= z + \sum_{j=1}^{m} G'_j(z)d_j \\
\mathcal{I}_2(z) &= z + d \\
\mathcal{I}_3(z) &= z
\end{aligned}
$$

- R and G'_j, for $j = 1, \ldots, m$, are new variables;

- the applied function is $id(a) = a$;

- the new domains are:

$$
\begin{aligned}
D_1 &= \{z + ld \mid z \in D, 0 \leq l \leq \bar{G}\} \\
D_{1,1} &= \{z + ld \mid z \in D, 0 \leq l < \bar{G}\} \\
D_{1,2} &= \{z + \bar{G}d \mid z \in D\}
\end{aligned}
$$

PROOF: In the proof, we use the condition $0 \leq G(z) \leq \bar{G}$ which is satisfied for all inputs.

For all $z \in D$, let $segm(z) = \{z + ld \mid 0 \leq l \leq \bar{G}\}$. By definition, for all $z' \in segm(z)$. $G'_j(z') = G_j(z)$. In fact, for all $z \in D$,

$$G'_j(z) = G'_j(z + d) = \ldots = G'_j(z + \bar{G}d)$$
$$= G^{tr}_j(z + \bar{G}d) = G_j(z).$$

Therefore, for all $z \in D$,

$$\mathcal{I}_1 \circ \mathcal{I}_0(z) = z + G(z)d + \sum_{j=1}^{m} G'_j(z + G(z)d)d_j$$

$$=_{[0 \leq G(z) \leq \bar{G}]} \quad z + G(z)d + \sum_{j=1}^{m} G_j(z)d_j = \mathcal{I}(z).$$

Finally, for all $z \in D$,

$$U(z) = R(\mathcal{I}_0(z))$$
$$= V(\mathcal{I}_1 \circ \mathcal{I}_0(z)) = V(\mathcal{I}(z)).$$

∎ 4.4.13

The second technique extends the scheme discussed for the previous technique to the case when the generator d is contained in $lin(D)$. It combines integral decomposition (see Proposition 3.2.21) with the redefinition of the coefficients $G_1, \ldots, G_m$. The technique is illustrated in Fig. 4.12 (in 3-dimensions) for two possible values of $\mathcal{I}(z)$ ($\mathcal{I}[in](z)$ in part a) and $\mathcal{I}[in'](z)$ in part b)).

Proposition 4.4.14 *[Decomposition 2]* Let $\mathcal{DD} = (D, U, V, \mathcal{I})$ be a finitely generated data dependence, with $\mathcal{I}(z) = z + G(z)d + \sum_{j=1}^{m} G_j(z)d_j$. Let G, Gj be non-negative and bounded over D and let $\bar{G}, \bar{G}_j$ be the least upper bounds of G, G_j, respectively. Let $d \in lin(D)$, P be the convex polyhedron generated by D and $\{d_1, \ldots, d_m\}$ and $dim(P) < n$. Consider $\pi \in P^{\perp}$, with $\pi \neq 0$, and the hyperplane $[\pi : 0]$ containing the domain D, and let $\hat{d} = d + \pi$. Let U and G_j be separable, for all j. Consider the system of equations $(DefS_{\{G_1, \ldots, G_m\}})^{tr, ren}$, defined by the renaming $ren(G_j) = G^{tr}_j$ and the translation $tr(z) = z + \bar{G}\hat{d}$.

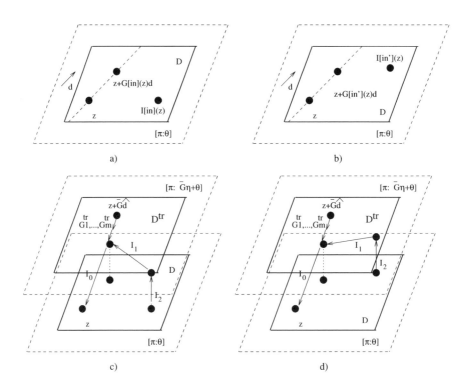

Fig. 4.13. Decomposition 2.

Then $\mathcal{DD}$ can be substituted by the data dependence

$$\mathcal{DD}' \;=\; (D, U, R^1, \mathcal{I}_0)$$

and the equations:

$$
\begin{aligned}
\mathbf{E}_1 &= (D_1, R^1, R^2, id, \mathcal{I}_1) \\
\mathbf{E}_2 &= (D_2, R^2, V, id, \mathcal{I}_2) \\
\mathbf{E}_{1_1} &= (D_{1,1}, G'_j, G'_1, id, \mathcal{I}_3) \\
\mathbf{E}_{1_2} &= (D_{1,2}, G'_j, G_1^{tr}, id, \mathcal{I}_4) \\
&\quad\cdots \\
\mathbf{E}_{m_1} &= (D_{1,1}, G'_j, G'_m, id, \mathcal{I}_3) \\
\mathbf{E}_{m_2} &= (D_{1,2}, G'_j, G_m^{tr}, id, \mathcal{I}_4) \\
\mathbf{E}_3 &= (D_2, G', in_{G'})
\end{aligned}
$$

where:

- the index mappings are:

$$
\begin{aligned}
\mathcal{I}_0(z) &= z + G(z)\hat{d} \\
\mathcal{I}_1(z) &= z + \sum_{j=1}^{m} G'_j(z)d_j \\
\mathcal{I}_2(z) &= z + G'(z)(-\pi) \\
\mathcal{I}_3(z) &= z + \hat{d} \\
\mathcal{I}_4(z) &= z
\end{aligned}
$$

- R^1, R^2, G' and G'_j, for $j = 1, \ldots, m$, are new variables;

- the applied functions are:

$$
\begin{aligned}
in_{G'}(z) &= (\pi \cdot z - \theta)/(\pi \cdot \hat{d}) \\
id(a) &= a
\end{aligned}
$$

- the new domains are:

$$
\begin{aligned}
D_1 &= \{z + l\hat{d} \mid z \in D, 0 \le l \le \bar{G}\} \\
D_{1,1} &= \{z + l\hat{d} \mid z \in D, 0 \le l < \bar{G}\} \\
D_{1,2} &= \{z + \bar{G}\hat{d} \mid z \in D\} \\
D_2 &= \{z + \sum_{j=1}^{m} l_j d_j \mid z \in D_1, 0 \le l_j \le \bar{G}_j\}
\end{aligned}
$$

PROOF: In the proof, we use the condition $0 \le G(z) \le \bar{G}$, which is satisfied for all inputs.

For all $z \in D$, let $segm(z) = \{z + ld \mid 0 \le l \le \bar{G}\}$. By definition, for all $z' \in segm(z)$, $G'_j(z') = G_j(z)$. In fact, for all $z \in D$,

$$G'_j(z) = G'_j(z + \hat{d}) = \ldots = G'_j(z + \bar{G}\hat{d})$$
$$= G^{tr}_j(z + \bar{G}\hat{d}) = G_j(z).$$

Therefore, for all $z \in D$,

$$\mathcal{I}_1 \circ \mathcal{I}_0(z) = z + G(z)\hat{d} + \sum_{j=1}^{m} G'_j(z + G(z)\hat{d})d_j$$

$$=_{[0 \le G(z) \le \bar{G}]} \quad z + G(z)\hat{d} + \sum_{j=1}^{m} G_j(z)d_j.$$

Also, as $\pi \in P^{\perp}$, then $\pi \cdot d_j = 0$ and $\pi \cdot (\sum_{j=1}^{m} G_j(z)d_j) = 0$. Hence, for all $z \in D$,

$$G'(\mathcal{I}_1 \circ \mathcal{I}_0(z)) = in_{G'}(\mathcal{I}_1 \circ \mathcal{I}_0(z))$$
$$= (\pi \cdot (\mathcal{I}_1 \circ \mathcal{I}_0(z)) - \theta)/(\pi \cdot \hat{d})$$
$$= (\pi \cdot (z + G(z)\hat{d} + \sum_{j=1}^{m} G_j(z)d_j) - \theta)/(\pi \cdot \hat{d})$$
$$= (\pi \cdot z + G(z)\pi \cdot \hat{d} + \pi \cdot (\sum_{j=1}^{m} G_j(z)d_j) - \theta)/(\pi \cdot \hat{d})$$
$$= (\theta + G(z)\pi \cdot \hat{d} + 0 - \theta)/(\pi \cdot \hat{d}) = G(z).$$

Therefore, for all $z \in D$,

$$\mathcal{I}_2 \circ \mathcal{I}_1 \circ \mathcal{I}_0(z) = \mathcal{I}_2(\mathcal{I}_1 \circ \mathcal{I}_0(z))$$
$$= \mathcal{I}_1 \circ \mathcal{I}_0(z) + G'(\mathcal{I}_1 \circ \mathcal{I}_0(z))(-\pi)$$
$$= \mathcal{I}_1 \circ \mathcal{I}_0(z) + G(z)(-\pi)$$
$$= z + G(z)\hat{d} + \sum_{j=1}^{m} G_j(z)d_j + G(z)(-\pi)$$
$$= z + G(z)(\pi + d - \pi) + \sum_{j=1}^{m} G_j(z)d_j$$
$$= z + G(z)d + \sum_{j=1}^{m} G_j(z)d_j = \mathcal{I}(z).$$

Finally, for all $z \in D$,

$$U(z) = R^1(\mathcal{I}_0(z)) = R^2(\mathcal{I}_1 \circ \mathcal{I}_0(z))$$
$$= V(\mathcal{I}_2 \circ \mathcal{I}_1 \circ \mathcal{I}_0(z)) = V(\mathcal{I}(z)).$$

■ 4.4.14

From the previous to results, we obtain the corollary:

Corollary 4.4.15   Let $\mathcal{DD} = (D, U, V, \mathcal{I})$ be a finitely generated data dependence, with $\mathcal{I}(z) = z + G(z)d + \sum_{j=1}^{m} G_j(z)d_j$. Let G be non-negative and bounded over D and $\bar{G}$ the least upper bound of G.

If $\bar{G} = 0$, then $\mathcal{DD}$ can be substituted by the data dependence:

$$\mathcal{DD}_0 \quad = \quad (D, U, V, \mathcal{I}_0)$$

where $\mathcal{I}_0(z) = z + \sum_{j=1}^{m} G_j(z)d_j$.

■ 4.4.15

4.5   Regularisation and Affine Scheduling

The preservation of affine scheduling by the regularisation techniques of the previous section can be proved by applying Propositions 3.3.1 and 3.3.2. The results are summarised in Table 4.1, where: the first column indicates each regularisation technique with, in brackets, the reference number of the corresponding formal statement; the generators of the embedding dependence cone before (C) and after (C') the application of each technique are given in the second and third columns, respectively. In particular, in such columns $r = -d$ and $r_j = -d_j$, for all $j = 1, \dots, m$, denote the generators of the corresponding dependence cone, while $\hat{r} = -\hat{d}, \check{r} = -\check{d}$ and $\rho = -\pi$, the new generators introduced by the various techniques; the last column indicates which of Propositions 3.3.1 and 3.3.2 applies. A line — indicates that no proposition applies and corresponds to a case in which the dependence cone is not modified by the regularisation technique.

Note that Propositions 3.3.1 and 3.3.2 apply to the dynamic case because of the assumption that the coefficients of a finitely generated index mapping are non-negative and bounded variables. In turn, this condition guarantees that for all inputs, the corresponding integral index mappings are characterised by coefficients which are also non-negative and bounded.

Technique	Generators of C	Generators of C'	Proposition
Unif. (4.4.6)	r	r	—
Unif. (4.4.8)	r	$r, \hat{r}, \check{r}$	3.3.1
Par. Unif. (4.4.11)	r	$r, \hat{r}, \check{r}$	3.3.1
Decomp. (4.4.13)	$r, r_1, \ldots, r_m$	$r, r_1, \ldots, r_m$	—
Decomp. (4.4.14)	$r, r_1, \ldots, r_m$	$r, r_1, \ldots, r_m, \hat{r}, \rho$	3.3.2

Table 4.1. Regularisation and the preservation of affine scheduling.

Technique	Condition	Extra Dimensions
Unif. (4.4.6)	$d \notin lin(D)$	no
Unif. (4.4.8)	$d \in lin(D)$ $dim(D) < n$	yes
Par. Unif. (4.4.11)	$d \in lin(D)$ $dim(D) < n$	yes
Decomp. (4.4.13)	$d \notin lin(D)$	no
Decomp. (4.4.14)	$d \in lin(D)$ $dim(P) < n$	yes

Table 4.2. Summary of regularisation techniques.

In order to guarantee the preservation of an affine scheduling for the whole specification we can consider (as we did for integral problems in Section 3.3) the dependence cone of the corresponding system of equations. Let **S** be such a system, and $C_{\mathbf{S}}$ its dependence cone. Then, when a regularisation technique is applied, the normal vector (π in the formal statements) to the hyperplane at the basis of the transformation can be chosen in the space $(C_{\mathbf{S}})^{\perp}$. This choice guarantees both that $\pi \in C^{\perp}$ and that affine schedulings are preserved for the system **S**.

4.6   Summary

The main objective of this chapter was to demonstrate that classes of dynamic problems can be transformed systematically into regular arrays in the context of classic synthesis methods. We have reached our objective by characterising a type of dynamic data dependencies (those which vary with the inputs of an algorithm) and identifying a subclass which can be represented and treated within Euclidean synthesis.

We based both the formalisation and treatment of this type of problems on a generalisation of integral index mappings, in which indexed variables replace integer functions as the coefficients of the mappings. We highlighted the implications of this substitution. In particular, we developed the notions of extended dependence graph and variable separability, with variable separability as the primary condition for the application of the regularisation techniques.

The regularisation techniques that we have developed are summarised in Table 4.2. In the table, the first column indicates the technique and in brackets the reference number of the corresponding formal result. The second column indicates the basic conditions for its applicability. In particular, d denotes a generator of the index mapping, D the domain of the data dependence, and P the polyhedron generated by d and D. The last column indicates whether the application of the technique may require an increase in the number of dimensions of the space.

A major difference with respect to integral techniques is the necessity of translating the subsystems of equations defining the coefficients of the index mappings in the computation space of the problem. This translation is necessary to reconfigure the data routing scheme through the initialisation of its control variables. This also limits the applicability of the techniques, and some interesting problems from the literature cannot be treated with our methods. One of these problems, that of Gaussian elimination with pivoting, will be discussed in the next chapter.

Chapter 5

Case Studies

In this chapter we will consider a number of problems from the literature in order to illustrate the application, and limitations, of the techniques which we have developed in the previous chapters.

The first problem we consider is a cyclic reduction algorithm for the solution of tridiagonal systems [Mod88, LoZa94]. A tridiagonal system is a system of linear equations whose matrix of coefficients is tridiagonal, i.e., the non-null entries of the matrix are all concentrated around the main diagonal in a band of width equal to 3. The solution of tridiagonal systems is at the base of several numerical applications, such as the solution of systems of partial differential equations [BuFa93]. Cyclic reduction algorithms for the solution of tridiagonal systems are known from the literature [Mod88, LoZa94]. In this chapter, we will show how the problem can be synthesised with our techniques through its specification as a system of integral recurrence equations.

The difficulties posed by integral data dependencies with unbounded coefficients will be discussed by considering the example of a decimation filter . Decimation filters are used for the conversion of the sampling rates of digital signals and constitute significant applications in digital signal processing [CrRa83]. We will show that the problem can be specified as a system of integral recurrence equations (the problem is, in particular, affine). However, because of the unbounded coefficients of its data dependencies, our techniques are not applicable. Instead, a simple reformulation of the problem will be given directly as a system of uniform recurrence equations. We will also show that not even affine regularisation techniques apply to the problem in its first specification.

Problem	Spec Type	Regularisation
Cyclic Reduction	Int.	Yes
M-to-1 Decimator	Int.	No (unbounded coefficient)
Knapsack	Dyn.	Yes
Gauss. Elim. with Pivoting	Dyn.	No (non-separable variables)

Table 5.1. Summary of the case studies.

As an example of a dynamic problem we will consider the well-studied knapsack problem[MaTo90, Hu82], an NP-hard combinatorial optimisation problem both of theoretical and practical interest. We will show that we can treat the knapsack problem as a dynamic finitely generated problem and that our techniques can be applied for the systematic derivation of corresponding regular array designs.

Finally, we will consider the technique of Gaussian elimination with partial pivoting [Mod88] [BaEl88, Ho-et-al89, ElBa90, Meg90] for the reduction of a dense matrix to a triangular form. The addition of pivoting to the elimination process increases its numerical stability and avoids breakdown due to division by zero. Pivoting, however, introduces dynamic data dependencies which cannot be handled by classic synthesis methods. When formulated as a dynamic problem, Gaussian elimination with pivoting is characterised by computations which cannot be separated. Therefore our techniques are not applicable. Instead, a static specification of the problem will be given, corresponding to regular array designs which have appeared in the literature [Meg90].

We have summarised the case studies in Table 5.1. The table indicates the problem, the type of the corresponding specification ("Int." stands for integral and "Dyn." for dynamic), whether our regularisation techniques apply and, if not, the condition that prevents their application.

5.1 Cyclic Reduction

The first problem that we consider is an algorithm for the solution of tridiagonal systems of equations through cyclic reduction. The following formulation of the problem is based on [Sa-et-al93]. Let $N = 2^r - 1$, for some integer r.

A tridiagonal set of N irreducible linear equations is:

$$
\begin{aligned}
a_{i-1}s_{i-2} + b_{i-1}x_{i-1} + c_{i-1}x_i &= d_{i-1} \\
a_i s_{i-1} + b_i x_i + c_i x_{i+1} &= d_i \\
a_{i+1}s_i + b_{i+1}x_{i+1} + c_{i+1}x_{i+2} &= d_{i+1}
\end{aligned}
$$

for $i = 2, 4, 6, \ldots, N - 1$, with non-null coefficients and boundary values $x_0 = x_{N+1} = 0$.

The solution of this set of equations with cyclic reduction consists of a so-called reduction phase followed by a backsubstitution phase. The reduction phase aims at reducing the system to a single equation in a single variable, whose value can be determined with a single arithmetic operation. The backsubstitution phase recursively substitutes the variable values already determined in the sets of equations generated in the reduction phase, until all variable values are determined.

The reduction phase includes $r - 1$ reduction steps. Let j denote the current step, with $j = 1, 2, \ldots, r - 1$, and let $g(j) = 2^{j-1}$. Each reduction step j consists of the computations of the following coefficients, for $i = 2g(j), 4g(j), \ldots, 2^r - 2g(j)$:

$$
a_i^j = \frac{-a_i^{j-1} a_{i-g(j)}^{j-1}}{b_{i-g(j)}^{j-1}}
$$

$$
b_i^j = b_i^{j-1} + \frac{-a_i^{j-1} c_{i-g(j)}^{j-1}}{b_{i-g(j)}^{j-1}} + \frac{-c_i^{j-1} a_{i-g(j)}^{j-1}}{b_{i+g(j)}^{j-1}}
$$

$$
c_i^j = \frac{-c_i^{j-1} c_{i+g(j)}^{j-1}}{b_{i+g(j)}^{j-1}}
$$

$$
d_i^j = d_i^{j-1} + \frac{-a_i^{j-1} d_{i-g(j)}^{j-1}}{b_{i-g(j)}^{j-1}} + \frac{-c_i^{j-1} d_{i+g(j)}^{j-1}}{b_{i+g(j)}^{j-1}}
$$

The initial values are $a_i^0 = a_i$, $b_i^0 = b_i$, $c_i^0 = c_i$ and $d_i^0 = d_i$.

At the end of the reduction phase, the remaining equation is:

$$
a_{2g(r-1)}^{r-1} x_0 + b_{2g(r-1)}^{r-1} x_{2g(r-1)} + c_{2g(r-1)}^{r-1} x_{4g(r-1)} = d_{2g(r-1)}^{r-1}.
$$

Note that $4g(r - 1) = 4 * 2^{r-2} = 2^r = N + 1$, hence $x_{4g(r-1)} = x_{N+1}$. Therefore, because of the boundary conditions $x_0 = x_{N+1} = 0$, the value of variable $x_{2g(r-1)}$ can be obtained from the above equation in a single operation.

The other values of x can be obtained in $r - 1$ backsubstitution steps. Once again, let j denote the current step, with $j = r - 1, r - 2, \ldots, 1$, and let $g'(j) = 2^j$. Then, at each step j, and for $i = g'(j), 3g'(j), \ldots, 2^r - g'(j)$, x_i can be determined as:

$$x_i \quad = \quad \frac{d_i^j - a_i^j x_{i-g'(j)} - c_i^j x_{i+g'(j)}}{b_i^j}$$

Note that the two phases are strictly sequential and can be treated separately.

5.1.1 Specification

Let us specify the two phases of the algorithm separately.

Reduction Phase

The formulation of the reduction phase in the previous section was that of a system of recurrence equations, with coefficients indexed by two indices i and j. We can specify this phase in $\mathbf{Z}^2$ by introducing the variables A, B, C and D, corresponding to the coefficients of the equations.

The main difficulty with this specification is that the computation points do not constitute convex polyhedral domains. We solve the problem by embedding the computation points in a convex polyhedral region D_1 and defining a control variable T to identify these computation points in D_1. The domain D_1 is defined as:

$$D_1 \quad = \quad \{(i,j) \mid 1 \le j \le r - 1, 2 \le i \le 2^r - 2\}$$

while the definition of T is based on the function:

$$f(i,j) = \begin{cases} 1 & i/2^{j-1} \in \{2, 4, 6, \ldots, 2^r - 2\} \\ 0 & \text{otherwise} \end{cases}$$

We make variables A, B, C and D undefined (equal to $\perp$) at each $(i,j) \in D_1$ such that $f(i,j) = 0$. The specification is:

(variable A)

$$\mathbf{E}_1 \quad = \quad (D_0, A, in_A)$$

$$\mathbf{E}_2 \;=\; (D_1, A, (T, A, A, B), f_1, (\mathcal{I}_0, \mathcal{I}_1, \mathcal{I}_2, \mathcal{I}_2))$$

(variable B)

$$\mathbf{E}_3 \;=\; (D_0, B, in_B)$$

$$\mathbf{E}_4 \;=\; (D_1, B, (T, B, A, C, B, C, A, B), f_2, (\mathcal{I}_0, \mathcal{I}_1, \mathcal{I}_1, \mathcal{I}_2, \mathcal{I}_2, \mathcal{I}_1, \mathcal{I}_2, \mathcal{I}_3))$$

(variable C)

$$\mathbf{E}_5 \;=\; (D_0, C, in_C)$$

$$\mathbf{E}_6 \;=\; (D_1, C, (T, C, C, B), f_1, (\mathcal{I}_0, \mathcal{I}_1, \mathcal{I}_3, \mathcal{I}_3))$$

(variable D)

$$\mathbf{E}_7 \;=\; (D_0, D, in_D)$$

$$\mathbf{E}_8 \;=\; (D_1, D, (T, D, A, D, B, C, D, B), f_2, (\mathcal{I}_0, \mathcal{I}_1, \mathcal{I}_1, \mathcal{I}_2, \mathcal{I}_2, \mathcal{I}_1, \mathcal{I}_3, \mathcal{I}_3))$$

(control variable T)

$$\mathbf{E}_9 \;=\; (D_1, T, f)$$

with:

- domains:

$$
\begin{aligned}
D_0 \;&=\; \{(i,j) \mid j = 0, 1 \le i \le 2^r - 1\} \\
D_1 \;&=\; \{(i,j) \mid 1 \le j \le r - 1, 2 \le i \le 2^r - 2\}
\end{aligned}
$$

- index mappings:

$$
\begin{aligned}
\mathcal{I}_0(i,j) \;&=\; (i,j) \\
\mathcal{I}_1(i,j) \;&=\; (i, j-1) \\
\mathcal{I}_2(i,j) \;&=\; (i - g(i,j), j-1) \\
\mathcal{I}_3(i,j) \;&=\; (i + g(i,j), j-1)
\end{aligned}
$$

where $g(i,j) = 2^{j-1}$; and

- applied functions:

$$
\begin{aligned}
in_A(i,j) \;&=\; a_i \\
in_B(i,j) \;&=\; b_i \\
in_C(i,j) \;&=\; c_i \\
in_D(i,j) \;&=\; d_i \\
f_1(t, a, b, c) \;&=\; \begin{cases} -a * b/c & t = 1 \\ \bot & \text{otherwise} \end{cases}
\end{aligned}
$$

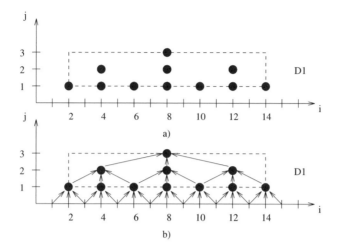

Fig. 5.1. Reduction phase: a) computation points; b) data dependence graph.

$$f_2(t, a, b, c, d, e, f, g) = \begin{cases} a - b * c/d - e * f/g & t = 1 \\ \perp & \text{otherwise} \end{cases}$$

$$f(i, j) = \begin{cases} 1 & i/2^{j-1} \in \{2, 4, 6, \dots, 2^r - 2\} \\ 0 & \text{otherwise} \end{cases}$$

A picture of the computation points of the phase is given in Fig. 5.1 a), for $r = 4$.

Backsubstitution Phase

The backsubstitution phase of the algorithm can also be formulated as a system of recurrences in $\mathbf{Z}^2$. To this end we introduce a variable X, corresponding to x, which we full index in $\mathbf{Z}^2$ (in the previous section x is indexed by a single index i). Let us consider the domains:

$$D_0' = \{(i, j) \mid j = r, 0 \le i \le 2^r\}$$
$$D_1' = \{(i, j) \mid 0 \le j \le r - 1, 0 \le i \le 2^r\}$$

and the control variable T' defined on D_1' by the function:

$$f'(i, j) = \begin{cases} 1 & 2^j/i \in \{1, 3, 7, \dots, 2^r - 1\} \\ 0 & \text{otherwise} \end{cases}$$

Then the backsubstitution phase may be specified as:

$$(\text{variable } X)$$
$$\mathbf{E}_1' \;=\; (D_0', X, in_x)$$
$$\mathbf{E}_2' \;=\; (D_1', X, (T', X, D, A, X, C, X, B), f, (\mathcal{I}_0, \mathcal{I}_1', \mathcal{I}_0, \mathcal{I}_0, \mathcal{I}_2', \mathcal{I}_0, \mathcal{I}_3', \mathcal{I}_0))$$
$$(\text{control variable})$$
$$\mathbf{E}_3' \;=\; (D_1', T', f')$$

with index mappings:

$$
\begin{aligned}
\mathcal{I}_0'(i,j) &= (i,j) \\
\mathcal{I}_1'(i,j) &= (i, j+1) \\
\mathcal{I}_2'(i,j) &= (i - g'(i,j), j+1) \\
\mathcal{I}_3'(i,j) &= (i + g'(i,j), j+1)
\end{aligned}
$$

where $g'(i,j) = 2^j$, and applied functions:

$$
\begin{aligned}
in_x(i,j) &= 0 \\[4pt]
f(t', a, b, c, d, e, f, g) &=
\begin{cases}
(b - c * d - e * f)/g & t' = 1 \\
a & \text{otherwise}
\end{cases} \\[4pt]
f'(i,j) &=
\begin{cases}
1 & 2^j/i \in \{1, 3, 7, \ldots, 2^r - 1\} \\
0 & \text{otherwise}
\end{cases}
\end{aligned}
$$

The values x_i correspond to the values $X(i, 0)$, for all $i = 0, 1, \ldots, 2^r$. The computation points of this phase are given in Fig. 5.2 a), for $r = 4$. In the figure, black nodes indicate points at which values x_i are actually computed, while white nodes the points at which already computed x_i are pipelined. Note the similarity existing between the two phases, as illustrated in Fig. 5.1 and Fig. 5.2.

5.1.2   Analysis of the Data Dependencies

Although several variables are involved, and the overall specification may appear rather complicated, only a few types of data dependencies characterise all the computations of the two phases. Besides, the two phases and their data dependencies are very similar. Therefore, for brevity, in the following discussion we restrict ourselves to the reduction phase of the algorithm.

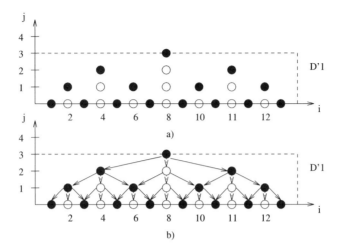

Fig. 5.2. Backsubstitution phase: a) computation points; b) data dependence graph.

Let U, V represent any of the variables A, B, C, D or T. All the data
dependencies of the phase reduce to one of the following cases:

$$
\begin{aligned}
\mathcal{DD}_0 &= (D_1, U, V, \mathcal{I}_0) \\
\mathcal{DD}_1 &= (D_1, U, V, \mathcal{I}_1) \\
\mathcal{DD}_2 &= (D_1, U, V, \mathcal{I}_2) \\
\mathcal{DD}_3 &= (D_1, U, V, \mathcal{I}_3)
\end{aligned}
$$

with index mappings:

$$
\begin{aligned}
\mathcal{I}_0(i, j) &= (i, j) \\
\mathcal{I}_1(i, j) &= (i, j - 1) \\
\mathcal{I}_2(i, j) &= (i - g(i, j), j - 1) \\
\mathcal{I}_3(i, j) &= (i + g(i, j), j - 1)
\end{aligned}
$$

where $g(i, j) = 2^{j-1}$. The data dependence graph is illustrated in Fig. 5.1 b).
The dependence cones $\Theta^*_{\mathcal{I}_1}$, $\Theta^*_{\mathcal{I}_2}$ and $\Theta^*_{\mathcal{I}_3}$ are illustrated in Fig. 5.3 a), b) and
c), respectively, and the overall dependence cone Θ^* in Fig. 5.4 a).
    The above index mappings are all integral and, in particular, $\mathcal{I}_0$ and $\mathcal{I}_1$
are uniform. An explicit integral form for the mappings is the following:

$$
\mathcal{I}_0(i, j) \;=\; (i, j)
$$

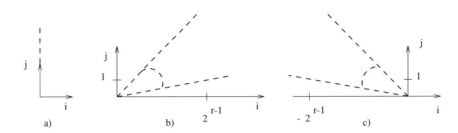

Fig. 5.3. Dependence cones: a) $\Theta^*_{\mathcal{I}_1}$; b) $\Theta^*_{\mathcal{I}_2}$; c) $\Theta^*_{\mathcal{I}_3}$.

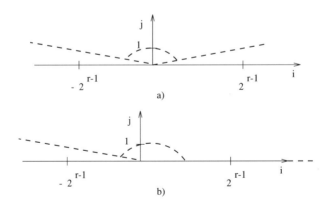

Fig. 5.4. Reduction phase: a) Θ^*; b) C.

$$
\begin{aligned}
\mathcal{I}_1(i,j) &= (i,j) + (0,-1) \\
\mathcal{I}_2(i,j) &= (i,j) + (-1,-1) + (g(i,j)-1)(-1,0) \\
\mathcal{I}_3(i,j) &= (i,j) + (2^{r-1},-1) + (2^{r-1}-g(i,j))(-1,0)
\end{aligned}
$$

The generators of the above index mappings have been chosen so that the cone C that they span (see Fig. 5.4 b)) contains the data dependence cone Θ^*, is pointed and has as extremal rays the vectors of the unimodular basis $\{(2^{r-1},-1),(1,0)\}$ (see also the discussion in Section 3.2.1). Moreover, the corresponding coefficients define non-negative and bounded integer functions over D_1. In particular, as for all $(i,j) \in D_1$, $1 \le g(i,j) = 2^{j-1} \le 2^{r-2}$, then:

$$
\begin{aligned}
0 &\le & g(i,j) - 1 & \le 2^{r-2} - 1 \\
2^{r-1} - 1 &\ge & 2^{r-1} - g(i,j) & \ge 2^{r-2}
\end{aligned}
$$

Although all the conditions are met for the application of integral regularisation techniques, array designs derived from this specification would be non-scalable and exhibit non-local (albeit regular) connections. To see that this is the case, let us consider the dependence cone more closely. As the size parameter r of the problem increases, the dependence cone Θ^* tends toward a non-pointed dependence cone. This is illustrated in Fig. 5.5 a), for $r = 3$ and $r = 4$. In addition, any possible decomposition of the index mappings is based on generators which are also dependent on the size parameter r, namely the vector $(-2^{r-1}, 1)$. (Fig. 5.5 b) shows how the cone C varies with r. By stepping ahead a little, Fig. 5.6 shows the data dependence graph that would result from a decomposition of the data dependencies according to the above formulation. Parts a) and b) of the figure illustrate the decomposition of $\mathcal{DD}_2$, while parts c) and d) that of $\mathcal{DD}_3$. The fact that vector $(-2^{r-1}, 1)$ depends on r implies that the data dependence graph corresponding to $\mathcal{DD}_3$ (hence the resulting signal flow graph under a space-time mapping) is not scalable, that is as r varies a different connection topology is required. If the array design is realised in hardware, this fact implies that an entirely new component has to be built any time we want to solve a problem of different size. On the contrary, a scalable design allows one simply to update an existing component, typically by adding or removing processing elements, when a problem of larger or smaller size needs to be addressed. Although scalability is a design constraint which is mainly relevant to hardware implementation, it is good design practice to try and generate fully scalable regular array designs. From a theoretically point of view, the notion of scalability can be captured by a notion of uniformity of the data dependencies with respect to

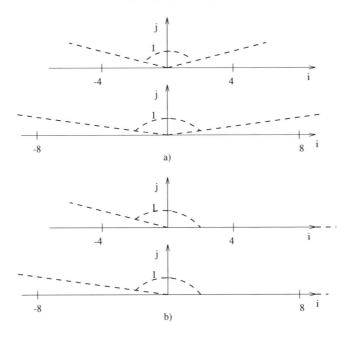

Fig. 5.5. Reduction phase: a) Θ^* for $r = 3$ and $r = 4$; b) C for $r = 3$ and $r = 4$.

the size parameters of the problem. This notion was developed by Quinton and Van Dongen in [QuVa89]. Fig. 5.7 a) and b) show the non-scaling data dependence graphs of $\mathcal{DD}_3$ for values of the size parameter $r = 3$ and $r = 4$, respectively. The reader may notice how the data dependence vectors change non uniformly as the size parameter increases, by comparing the sub-graph in b) corresponding to the graph in a).

5.1.3   New Specification

We want to reformulate the specification so that the data dependence graph for its reduction phase assumes the form of Fig. 5.8. As we will see this new specification will result in scalable regular array designs. The transformation we are looking for is defined by the mapping $\mathcal{T}(i, j) = (i + 2^j - 1, j)$ in $\mathbf{Z}^2$. The computation domain D_1 becomes

$$D_1 = \{(i, j) \mid 1 \le j \le r - 1, 3 \le i \le 2^r - 1\}$$

and the actual computations in D_1 are identified by the guard:

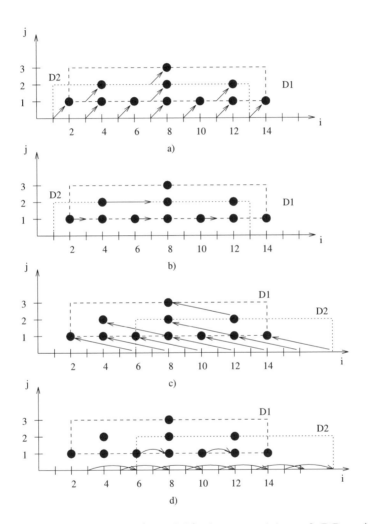

Fig. 5.6. Reduction phase: a) and b) decomposition of $\mathcal{DD}_2$; c) and d) decomposition of $\mathcal{DD}_3$.

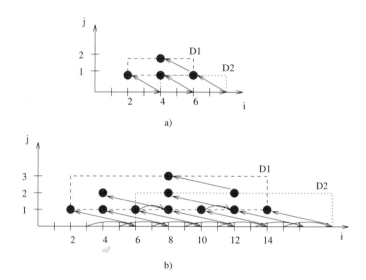

Fig. 5.7. Data dependence graph of $\mathcal{DD}_3$ for: a) $r = 3$; b) $r = 4$.

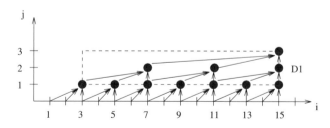

Fig. 5.8. Reduction phase (new specification): data dependence graph.

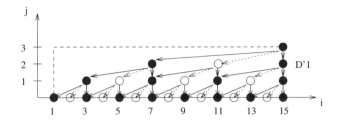

Fig. 5.9. Backsubstitution phase (new specification): data dependence graph.

$$f(i,j) = \begin{cases} 1 & (i - 2^j + 1)/2^{j-1} \in \{2, 4, 6, \ldots, 2^r - 2\} \\ 0 & \text{otherwise} \end{cases}$$

The new index mappings are:

$$\begin{aligned}
\mathcal{I}_0(i,j) &= (i,j) \\
\mathcal{I}_1(i,j) &= (i - g(i,j), j - 1) \\
\mathcal{I}_2(i,j) &= (i - g'(i,j), j - 1) \\
\mathcal{I}_3(i,j) &= (i, j - 1)
\end{aligned}$$

where $g(i,j) = 2^{j-1}$ and $g'(i,j) = 2^j$. The input domain D_0 and the applied functions are the same as in the previous specification.

A similar transformation applied to the backsubstitution phase produces the data dependence graph of Fig. 5.9.

5.1.4 Analysis of the New Data Dependencies

Once again let us restrict ourselves to the reduction phase of the algorithm, and let U, V represent any of the variables A, B, C, D or T. All data dependencies reduce to one of the following cases:

$$\begin{aligned}
\mathcal{DD}_0 &= (D_1, U, V, \mathcal{I}_0) \\
\mathcal{DD}_1 &= (D_1, U, V, \mathcal{I}_1) \\
\mathcal{DD}_2 &= (D_1, U, V, \mathcal{I}_2) \\
\mathcal{DD}_3 &= (D_1, U, V, \mathcal{I}_3)
\end{aligned}$$

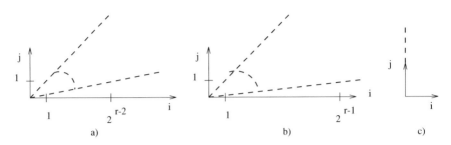

Fig. 5.10. Dependence cones relative to: a) $\mathcal{I}_1$; b) $\mathcal{I}_2$; c) $\mathcal{I}_3$.

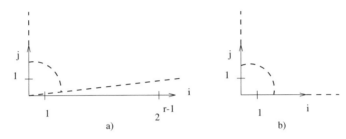

Fig. 5.11. Reduction phase (new specification): a) Θ^*; b) C.

with index mappings:

$$\begin{aligned}
\mathcal{I}_0(i,j) &= (i,j) \\
\mathcal{I}_1(i,j) &= (i - g(i,j), j - 1) \\
\mathcal{I}_2(i,j) &= (i - g'(i,j), j - 1) \\
\mathcal{I}_3(i,j) &= (i, j - 1)
\end{aligned}$$

where $g(i,j) = 2^{j-1}$ and $g'(i,j) = 2^j$. The dependence cones relative to $\mathcal{I}_1, \mathcal{I}_2$, and $\mathcal{I}_3$ are sketched in Fig. 5.10 a), b) and c), respectively, while the overall dependence cone Θ^* is illustrated in Fig. 5.11 a).

Explicit integral forms of the index mappings are:

$$\begin{aligned}
\mathcal{I}_0(i,j) &= (i,j) \\
\mathcal{I}_1(i,j) &= (i,j) + (-1,-1) + (g(i,j) - 1)(-1,0) \\
\mathcal{I}_2(i,j) &= (i,j) + (-1,-1) + (g'(i,j) - 1)(-1,0) \\
\mathcal{I}_3(i,j) &= (i,j) + (0,-1)
\end{aligned}$$

The cone C generated by $(0,1), (1,0)$ and $(1,1)$ is pointed and contains Θ^* (see Fig. 5.11 b)). Also, all the coefficients of the mappings define non-negative and bounded integer functions over D_1. In particular, for all $(i,j) \in$

D_1, $1 \le g(i,j) = 2^{j-1} \le 2^{r-2}$, $2 \le g'(i,j) = 2^j \le 2^{r-1}$, and:

$$0 \le \quad g(i,j) - 1 \quad \le 2^{r-2} - 1$$
$$1 \le \quad g'(i,j) - 1 \quad \le 2^{r-1} - 1$$

5.1.5 Regularisation

Both integral data dependencies $\mathcal{DD}_1$ and $\mathcal{DD}_2$ are not atomic. Therefore their decomposition is necessary before uniformisation may be applied. We present the two transformations separately.

Decomposition

Both data dependencies admit a simple decomposition, as their first component defines a uniform (hence, linear and injective) index mapping for which an inverse can be determined directly.

We substitute $\mathcal{DD}_1$ with the atomic integral data dependencies:

$$\mathcal{DD}'_1 = (D_1, U, P, \mathcal{I}'_1)$$
$$\mathcal{DD}_4 = (D_2, P, V, \mathcal{I}_4)$$

where P is a new variable, the index mappings are:

$$\mathcal{I}'_1(i,j) = (i,j) + (-1,-1)$$
$$\mathcal{I}_4(i,j) = (i,j) + (g'(i,j) - 1)(-1,0)$$

with $g'(i,j) = 2^j$, and the new domain is:

$$D_2 = \{(i-1, j-1) \mid (i,j) \in D_1\} = \{(i,j) \mid 0 \le j \le r-2, 2 \le i \le 2^r - 2\}.$$

Similarly, $\mathcal{DD}_2$ can be substituted by:

$$\mathcal{DD}'_2 = (D_1, U, Q, \mathcal{I}'_1)$$
$$\mathcal{DD}_5 = (D_2, Q, V, \mathcal{I}_5)$$

with index mappings:

$$\mathcal{I}'_1(i,j) = (i,j) + (-1,-1)$$
$$\mathcal{I}_5(i,j) = (i,j) + (g''(i,j) - 1)(-1,0)$$

where $g''(i,j) = 2^{j+1}$.

The effects of these substitutions are illustrated in Fig. 5.12 a) and b) for the components of $\mathcal{DD}_1$ and Fig. 5.12 c) and d) for the components of $\mathcal{DD}_2$.

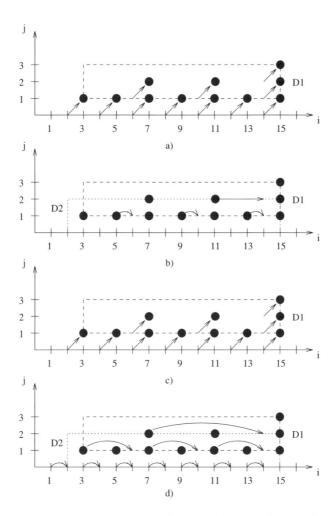

Fig. 5.12. Decomposition of: a) and b) $\mathcal{DD}_2$; a) and c) $\mathcal{DD}_2$.

Uniformisation

We can now proceed to the uniformisation of $\mathcal{DD}_4$ and $\mathcal{DD}_5$, the only remaining non-uniform data dependencies.

The two cases are similar. In particular, by adopting the notation of Section 3.2.3, in both cases $d = (-1, 0)$. Also $lin(D_2) = \mathbf{Z}^2$ and d is contained in this space. Hence, in both cases, a reindexing in $\mathbf{Z}^3$ is required. We choose a new index k and a system of axes i, j, k in this order. With this choice, a hyperplane $[\pi : \theta]$ containing the domain D_2 is determined by the vector $\pi = (0, 0, 1)$ and the coefficient $\theta = 0$. Hence, the uniformisation directions are $\hat{d} = d + \pi = (-1, 0, 1)$ and $\check{d} = d - \pi = (-1, 0, -1)$.

The differences between the two cases are: the routing domains, as the upper bounds of the coefficients of the index mappings are different; and the values of the control variables, as the coefficients of the index mappings define different functions on D_2.

The data dependence $\mathcal{DD}_4$ can be replaced by:

$$\mathcal{DD}'_4 = (D_2, P, R^1, \mathcal{I}_0)$$

where routing and control variables are defined by the equations:

$$
\begin{aligned}
&\text{(routing variables } R^1, R^2 \text{)}\\
\mathbf{E}_{10} &= (D_{2_1}, R^1, (\alpha, \beta, \gamma, R^1, R^2, R^2), f_3, (\mathcal{I}_0, \mathcal{I}_0, \mathcal{I}_0, \mathcal{I}'_4, \mathcal{I}_0, \mathcal{I}_6))\\
\mathbf{E}_{11} &= (D_{2_{2,1}}, R^2, R^2, id, \mathcal{I}_7)\\
\mathbf{E}_{12} &= (D_{2_{2,2}}, R^2, V, id, \mathcal{I}_0)\\
&\text{(control variables } \alpha, \beta \text{ and } \gamma)\\
\mathbf{E}_{13} &= (D_{2_{1,1}}, \alpha, \alpha, id, \mathcal{I}'_4)\\
\mathbf{E}_{14} &= (D_{2_{1,2}}, \alpha, in_\alpha)\\
\mathbf{E}_{15} &= (D_{2_{1,1}}, \beta, \beta, id, \mathcal{I}'_4)\\
\mathbf{E}_{16} &= (D_{2_{1,2}}, \beta, in_\beta)\\
\mathbf{E}_{17} &= (D_{2_{1,1}}, \gamma, \gamma, dec, \mathcal{I}'_4)\\
\mathbf{E}_{18} &= (D_{2_{1,2}}, \gamma, in_\gamma)
\end{aligned}
$$

with:

- index mappings:

$$
\begin{aligned}
\mathcal{I}_0(i, j, k) &= (i, j, k)\\
\mathcal{I}'_4(i, j, k) &= (i, j, k) + (-1, 0, 1)
\end{aligned}
$$

$$\begin{aligned}
\mathcal{I}_6(i,j,k) &= (i,j,k) + (-1,0,0) \\
\mathcal{I}_7(i,j,k) &= (i,j,k) + (-1,0,-1)
\end{aligned}$$

- applied functions (where $\bar{g}' = \lfloor (2^{r-2} - 1)/2 \rfloor$):

$$\begin{aligned}
in_\alpha(i,j,k) &= \lfloor (g'(i + \bar{g}', j, k - \bar{g}') - 1)/2 \rfloor \\
in_\beta(i,j,k) &= g'(i + \bar{g}', j, k - \bar{g}') \bmod 2 \\
in_\gamma(i,j,k) &= \bar{g}' \\
id(a) &= a \\
dec(a) &= a - 1 \\
f_3(\alpha, \beta, \gamma, a, b, c) &= \begin{cases} a & \alpha \neq \gamma \\ b & \alpha = \gamma, \beta = 0 \\ c & \alpha = \gamma, \beta = 1 \end{cases}
\end{aligned}$$

- domains:

$$\begin{aligned}
D_{2_1} &= \{(i,j,k) \mid 1 - k \leq i \leq 2^r - 2 - k, 0 \leq k \leq \bar{g}', \\
&\quad\ 0 \leq j \leq r - 2\} \\
D_{2_{1,1}} &= \{(i,j,k) \in D_{2_1} \mid k < \bar{g}'\} \\
D_{2_{1,2}} &= \{(i,j,k) \in D_{2_1} \mid k = \bar{g}'\} \\
D_{2_2} &= \{(i,j,k) \mid k - 2\bar{g}' \leq i \leq 2^r - 2 - k, 0 \leq k \leq \bar{g}', \\
&\quad\ 0 \leq j \leq r - 2\} \\
D_{2_{2,1}} &= \{(i,j,k) \in D_{2_2} \mid k > 0\} \\
D_{2_{2,2}} &= \{(i,j,k) \in D_{2_2} \mid k = 0\}.
\end{aligned}$$

Similarly, the uniformisation of $\mathcal{DD}_5$ yields the data dependence:

$$\mathcal{DD}'_5 = (D_2, Q, S^1, \mathcal{I}_0)$$

and the equations:

(routing variables S^1, S^2)

$$\begin{aligned}
\mathbf{E}_{19} &= (D_{2_3}, S^1, (\xi, \phi, \psi, S^1, S^2, S^2), f_3, (\mathcal{I}_0, \mathcal{I}_0, \mathcal{I}_0, \mathcal{I}'_4, \mathcal{I}_0, \mathcal{I}_6)) \\
\mathbf{E}_{20} &= (D_{2_{4,1}}, S^2, S^2, id, \mathcal{I}_7) \\
\mathbf{E}_{21} &= (D_{2_{4,2}}, S^2, V, id, \mathcal{I}_0)
\end{aligned}$$

(control variables ξ, ϕ and ψ)

$$\mathbf{E}_{22} = (D_{23,1}, \xi, \xi, id, \mathcal{I}_4')$$
$$\mathbf{E}_{23} = (D_{23,2}, \xi, in_\xi)$$
$$\mathbf{E}_{24} = (D_{23,1}, \phi, \phi, id, \mathcal{I}_4')$$
$$\mathbf{E}_{25} = (D_{23,2}, \phi, in_\phi)$$
$$\mathbf{E}_{26} = (D_{23,1}, \psi, \psi, dec, \mathcal{I}_4')$$
$$\mathbf{E}_{27} = (D_{23,2}, \psi, in_\psi)$$

where:

- the index mappings are:

$$\mathcal{I}_0(i, j, k) = (i, j, k)$$
$$\mathcal{I}_4'(i, j, k) = (i, j, k) + (-1, 0, 1)$$
$$\mathcal{I}_6(i, j, k) = (i, j, k) + (-1, 0, 0)$$
$$\mathcal{I}_7(i, j, k) = (i, j, k) + (-1, 0, -1)$$

- the applied functions are (where $\bar{g}'' = \lfloor (2^{r-1} - 1)/2 \rfloor$):

$$in_\xi(i, j, k) = \lfloor (g''(i + \bar{g}'', j, k - \bar{g}'') - 1)/2 \rfloor$$
$$in_\phi(i, j, k) = (g''(i + \bar{g}'', j, k - \bar{g}'') - 1) \bmod 2$$
$$in_\psi(i, j, k) = \bar{g}''$$
$$id(a) = a$$
$$dec(a) = a - 1$$
$$f_3(\xi, \phi, \psi, a, b, c) = \begin{cases} a & \xi \neq \psi \\ b & \xi = \psi, \phi = 0 \\ c & \xi = \psi, \phi = 1 \end{cases}$$

- the domains are:

$$D_{23} = \{(i, j, k) \mid 1 - k \leq i \leq 2^r - 2 - k, 0 \leq k \leq \bar{g}'',$$
$$0 \leq j \leq r - 2\}$$
$$D_{23,1} = \{(i, j, k) \in D_{23} \mid k < \bar{g}''\}$$
$$D_{23,2} = \{(i, j, k) \in D_{23} \mid k = \bar{g}''\}$$
$$D_{24} = \{(i, j, k) \mid k - 2\bar{g}'' \leq i \leq 2^r - 2 - k, 0 \leq k \leq \bar{g}'',$$
$$0 \leq j \leq r - 2\}$$
$$D_{24,1} = \{(i, j, k) \in D_{24} \mid k > 0\}$$
$$D_{24,2} = \{(i, j, k) \in D_{24} \mid k = 0\}.$$

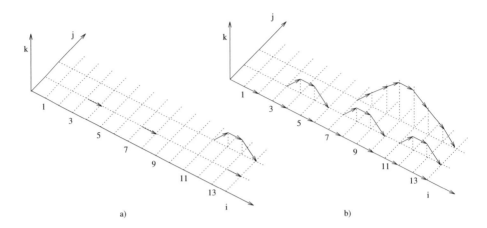

Fig. 5.13. Uniformisation of: a) $\mathcal{DD}_4$; b) $\mathcal{DD}_5$.

The data dependence graphs corresponding to $\mathcal{DD}'_4$ and $\mathcal{DD}'_5$ are sketched in Fig. 5.13 a) and b), respectively, where only the routing paths of interest are illustrated.

5.1.6   Space-Time Mapping

A uniform system of equations is obtained by applying regularisation techniques to all the data dependencies of the specification as explained in the previous section. The resulting uniform specification includes 35 variables, which we have summarised in Table 5.2.

The uniform data dependencies of the specification are summarised in Tables 5.3–5.7, where: the first column indicates the pair of variables which are related; the second and third columns, respectively, the new data dependencies introduced by decomposition and uniformisation (when applicable, a dash, –, otherwise); and the last column the corresponding (uniform) data dependence vector.

By considering all data dependence vectors, the resulting dependence cone is

$$\Theta^* = cone(\{(0,1,0),(1,1,0),(1,0,0),(1,0,-1),(1,0,1)\}),$$

with extremal rays $(0,1,0),(1,0,-1)$ and $(1,0,1)$. The cone is illustrated in Fig. 5.14.

Such a cone is pointed and, according to the conditions discussed in Section 2.2.6, an affine timing function for the specification is determined by

Type	Name	Number
Original variables	A, B, C, D	4
Original control variables	T	1
Decomposition variables	P_A, P_B, P_C, P_D Q_A, Q_B, Q_C, Q_D	8
Uniformisation variables	$R_A^1, R_B^1, R_C^1, R_D^1$ $R_A^2, R_B^2, R_C^2, R_D^2$ $S_A^1, S_B^1, S_C^1, S_D^1$ $S_A^2, S_B^2, S_C^2, S_D^2$	16
Uniformisation control variables	α, β, γ ξ, ψ, ϕ	6

Table 5.2. Variables in the final specification.

Pair	Decomp.	Unifor.	DD Vector
A, T	–	–	$(0, 0, 0)$
A, A	–	–	$(0, 1, 0)$
A, A	A, P_A	–	$(1, 1, 0)$
	P_A, A	P_A, R_A^1	$(0, 0, 0)$
		R_A^1, R_A^1	$(1, 0, -1)$
		R_A^1, R_A^2	$(0, 0, 0)$
		R_A^1, R_A^1	$(1, 0, 0)$
		R_A^2, R_A^2	$(1, 0, 1)$
		R_A^2, A	$(0, 0, 0)$
A, B	A, P_B	–	$(1, 1, 0)$
	P_B, B	P_B, R_B^1	$(0, 0, 0)$
		R_B^1, R_B^1	$(1, 0, -1)$
		R_B^1, R_B^2	$(0, 0, 0)$
		R_B^1, R_B^1	$(1, 0, 0)$
		R_B^2, R_B^2	$(1, 0, 1)$
		R_B^2, B	$(0, 0, 0)$

Table 5.3. Data dependencies related to variable A.

$Pair$	$Decomp.$	$Unifor.$	$DDVector$
B,T	$-$	$-$	$(0,0,0)$
B,B	$-$	$-$	$(0,1,0)$
B,A	$-$	$-$	$(0,1,0)$
B,C	B,P_C	$-$	$(1,1,0)$
	P_C,C	P_C,R^1_C	$(0,0,0)$
		R^1_C,R^1_C	$(1,0,-1)$
		R^1_C,R^2_C	$(0,0,0)$
		R^1_C,R^1_C	$(1,0,0)$
		R^2_C,R^2_C	$(1,0,1)$
		R^2_C,C	$(0,0,0)$
B,B	B,P_B	$-$	$(1,1,0)$
	P_B,B	see Table 5.3	see Table 5.3
B,C	$-$	$-$	$(0,1,0)$
B,A	B,P_A	$-$	$(1,1,0)$
	P_A,A	see Table 5.3	see Table 5.3
B,B	B,Q_B	$-$	$(1,1,0)$
	Q_B,B	Q_B,S^1_B	$(0,0,0)$
		S^1_B,S^1_B	$(1,0,-1)$
		S^1_B,S^2_B	$(0,0,0)$
		S^1_B,S^1_B	$(1,0,0)$
		S^2_B,S^2_B	$(1,0,1)$
		S^2_B,B	$(0,0,0)$

Table 5.4. Data dependencies related to variable B.

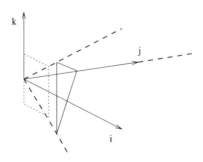

Fig. 5.14. Dependence cone Θ^* after uniformisation.

Pair	Decomp.	Unifor.	DDVector
C,T	$-$	$-$	$(0,0,0)$
C,C	$-$	$-$	$(0,1,0)$
C,C	C,Q_C	$-$	$(1,1,0)$
	Q_C,C	Q_C, S_C^1	$(0,0,0)$
		S_C^1, S_C^1	$(1,0,-1)$
		S_C^1, S_C^2	$(0,0,0)$
		S_C^1, S_C^1	$(1,0,0)$
		S_C^2, S_C^2	$(1,0,1)$
		S_C^2, C	$(0,0,0)$
C,B	C,Q_B	$-$	$(1,1,0)$
	Q_B,B	see Table 5.4	see Table 5.4

Table 5.5. Data dependencies relative to variable C.

any vector $\lambda = (\lambda_1, \lambda_2, \lambda_3)$ in $\mathbf{Z}^3$, such that:

$$\lambda_2 > 0$$
$$\lambda_1 - \lambda_3 > 0$$
$$\lambda_1 + \lambda_3 > 0.$$

Note that, according to our discussion in Section 2.2.5, further inequalities may be derived by considering the properties of time optimality and non-negativeness of the corresponding timing functions. In particular, such inequalities can be obtained by considering the generators (vertices and rays) of the computation domain of the specification, which is the smallest convex polyhedral set containing all the computation domains of the system. The resulting inequalities define an integer linear programming problem whose solution provides an optimal affine scheduling. For brevity, here (and in the following examples) we omit the complete formulation of the problem. The reader may be convinced that the vector $\lambda = (1,1,0)$ provides such an optimal solution, corresponding to the affine timing function $t(i,j,k) = i+j$. To make t non-negative, a suitable delay can be added so that t associates an initial time 0 with the first set of computations of the algorithm. With this scheduling the algorithm executes in $O(N + 2\lfloor N/2 \rfloor + 2)$ time steps, where $N = 2^r - 1$.

A possible projection vector is any $u = (u_1, u_2, u_3)$ in $\mathbf{Z}^3$ such that $\lambda \cdot u \neq 0$, i.e., $u_1 + u_2 \neq 0$. For instance, $u = (1,0,0)$ satisfies the requirement

Pair	Decomp.	Unifor.	DDVector
D,T	$-$	$-$	$(0,0,0)$
D,D	$-$	$-$	$(0,1,0)$
D,A	$-$	$-$	$(0,1,0)$
D,D	D,P_D	$-$	$(1,1,0)$
	P_D,D	P_D,R_D^1	$(0,0,0)$
		R_D^1,R_D^1	$(1,0,-1)$
		R_D^1,R_D^2	$(0,0,0)$
		R_D^1,R_D^1	$(1,0,0)$
		R_D^2,R_D^2	$(1,0,1)$
		R_D^2,D	$(0,0,0)$
D,B	D,P_B	$-$	$(1,1,0)$
	P_B,B	see Table 5.3	see Table 5.3
D,C	$-$	$-$	$(0,1,0)$
B,A	B,P_A	$-$	$(1,1,0)$
	P_A,A	see Table 5.3	see Table 5.3
D,D	D,Q_D	$-$	$(1,1,0)$
	Q_D,D	Q_D,S_D^1	$(0,0,0)$
		S_D^1,S_D^1	$(1,0,-1)$
		S_D^1,S_D^2	$(0,0,0)$
		S_D^1,S_D^1	$(1,0,0)$
		S_D^2,S_D^2	$(1,0,1)$
		S_D^2,D	$(0,0,0)$
D,B	D,Q_B	$-$	$(1,1,0)$
	Q_B,B	see Table 5.4	see Table 5.4

Table 5.6. Data dependencies relative to variable D.

Pair	Decomp.	Unifor.	DDVector
α,α	$-$	$-$	$(1,0,-1)$
β,β	$-$	$-$	$(1,0,-1)$
γ,γ	$-$	$-$	$(1,0,-1)$
ξ,ξ	$-$	$-$	$(1,0,-1)$
ϕ,ϕ	$-$	$-$	$(1,0,-1)$
ψ,ψ	$-$	$-$	$(1,0,-1)$

Table 5.7. Data dependencies relative to the routing control variables.

and corresponds to the allocation function $a(i, j, k) = (j, k)$. The effect of the space-time mapping $[t, a]$ on the (uniform) data dependence vectors of the specification is summarised in Table 5.8, and the image of the data dependence graph under $[t, a]$ is the signal flow graph of Fig. 5.15 a). In the figure, two types of nodes are indicated, which correspond to two basic types of processing elements (see later on). For simplicity, in the figure, unit arc labels have been omitted. Note that here we have considered only a single instance of each data dependence vector. Hence, this signal flow graph is a simplified version of the actual signal flow graph of the specification, in which each arc should be replicated for all the corresponding data dependencies of the system.

The signal flow graph together with the information provided by the recurrences are used for a detailed description of the array design. In particular, a processing element is associated with each node of the signal flow graph and a communication channel or a memory cell with each of its arcs. The operations at each processing element are specified by the applied functions of the recurrences whose computation points are mapped onto that processing element by the allocation function. The association of variables and communication channels or memory cells is also determined by the allocation function though the mapping of the corresponding data dependence vectors. A description of the two basic cells of the regular array corresponding to the signal flow graph in Fig. 5.15 a) is given in Fig. 5.16 a). In the figure we have assumed that: $U, V \in \{A, B, C, D\}$; $P, Q \in \{P_A, Q_A, P_B, Q_B, P_C, Q_C, P_D, Q_D\}$; $R^1, S^1 \in \{R_A^1, S_A^1, R_B^1, S_B^1, R_C^1, S_C^1, R_D^1, S_D^1\}$; and $R^2, S^2 \in \{R_A^2, S_A^2, R_B^2, S_B^2, R_C^2, S_C^2, R_D^2, S_D^2\}$. Dotted arrows correspond to control signals. Note that the loops of the signal flow graph correspond to memory cells in the processing elements. A slash, /, indicates an alternative (between variables of the two routing schemes of Section 5.1.5, while a single quote, ', denotes an output signal. Note that the description of the processing elements is only partial, as a complete description should include signals corresponding to all 35 variables of the specification.

Other compatible projection vectors for the given timing function t are, for instance, $u' = (1, 1, 0)$ and $u'' = (0, 1, 0)$, corresponding to the allocation functions $a'(i, j, k) = (i - j, k)$ and $a''(i, j, k) = (i, k)$, respectively. The effect of the resulting space-time mappings on the data dependence vectors are also given in Table 5.8, while Fig. 5.15 b) and c) illustrate the corresponding signal flow graphs, and Fig. 5.16 b) and c) the corresponding basic processing elements.

d	$\lambda \cdot d$	$a(d)$	$a'(d)$	$a''(d)$
$(0,1,0)$	1	$(1,0)$	$(-1,0)$	$(0,0)$
$(1,1,0)$	2	$(1,0)$	$(0,0)$	$(1,0)$
$(1,0,0)$	1	$(0,0)$	$(1,0)$	$(1,0)$
$(1,0,-1)$	1	$(0,-1)$	$(1,-1)$	$(1,-1)$
$(1,0,1)$	1	$(0,1)$	$(1,1)$	$(1,1)$

Table 5.8.   Transformation of the data dependence vectors under the space-time mappings $[t,a]$, $[t,a']$, and $[t,a'']$.

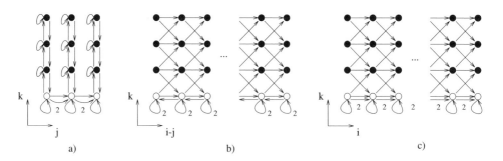

Fig. 5.15. Signal flow graphs for: a) $[t,a]$; b) $[t,a']$; c) $[t,a'']$.

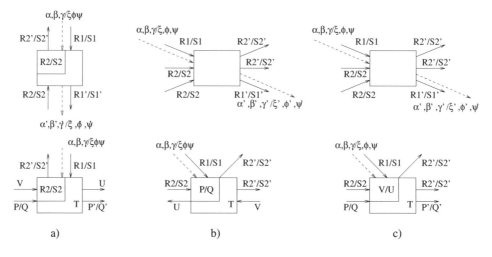

Fig. 5.16. Basic processing elements corresponding to: a) $[t,a]$; b) $[t,a']$; c) $[t,a'']$.

5.2 N Points FIR Filter for M-to-1 Decimation

One of the most fundamental concepts of digital signal processing [CrRa83] is the idea of sampling a continuous process to obtain a set of numbers which, in some sense, is representative of the characteristics of the process being sampled. Let $x_C(t)$ denote a continuous function of the process being sampled, where t is a continuous variable (typically time) and $-\infty < t < \infty$. A set of samples of x_C can be defined by another function $x_D(n)$, with $-\infty < n < \infty$, where the correspondence between t and n can be expressed by an equation $n = q(t)$, for some function q specified by the sampling process.

A common form of sampling, called uniform (periodic) sampling, is one in which $q(t) = t/T = n$, where n is an integer. That is, the samples $x_D(n)$ are uniformly spaced (occurring T apart) in the dimension t. T is called the sampling period.

In some cases the input signal may already be sampled at some predetermined sampling period T and the goal is to convert this sampled signal into a new sampled signal with a different sampling period T'. If $T' > T$ this digital conversion is called decimation. In particular, an M-to-1 decimator, for some natural number M, is a decimator which discards $M - 1$ every M (output) samples.

Decimation is usually obtained by filtering the input digital signal. A Finite Impulse Response (FIR) filter is a digital filter whose impulse response $h(k)$ is of finite duration, i.e., it is zero outside a finite interval of samples k. In particular, the direct form of an N points FIR filter is the convolution

$$y(m) = \sum_{n=0}^{N-1} h(n)x(m - n)$$

where the filter response h is assumed to be 0 for $n < 0$ and $n > N - 1$.

By combining the above definitions, the direct form of an N points FIR filter for M-to-1 decimation corresponds to the convolution:

$$y(m) = \sum_{n=0}^{N-1} h(n)x(Mm - n)$$

for $m \geq 0$.

5.2.1   Specification

We want to specify the problem as a system of integral recurrence equations.
The first step is to transform equation

$$y(m) = \sum_{n=0}^{N-1} h(n)x(Mm - n)$$

by exploiting the associativity of addition and introducing a variable Y which
accumulates the partial sums. Then each $y(m)$ can be computed by the
system of equations:

$$
\begin{aligned}
Y(m,0) &= 0 \\
Y(m,1) &= Y(m,0) + h(0)x(Mm) \\
Y(m,2) &= Y(m,1) + h(1)x(Mm - 1) \\
&\cdots \\
Y(m,N) &= Y(m,N-1) + h(N-1)x(Mm - N + 1)
\end{aligned}
$$

That is, more concisely:

$$
\begin{aligned}
Y(m,0) &= 0 \\
Y(m,j) &= Y(m,j-1) + h(j-1)x(Mm - j + 1)
\end{aligned}
$$

for $j = 1, \ldots, N$.

By introducing two new variables H and X, corresponding to h and x,
respectively, we can specify the problem in 2 dimensions as follows. An index
i is used which corresponds to m. For simplicity, let us assume that the new
digital signal y is sampled starting at $i = 1$, so that y is computed for $i \geq 1$.
Variable X is initialised to the values of x, for all i, while variable H is
initialised to the values of h. As those values are used in each convolution,
variable H also pipelines the values through the computation space. The
system of equations is the following:

$$
\begin{aligned}
\mathbf{E}_1 &= (D_1, X, in_X) \\
\mathbf{E}_2 &= (D_2, Y, in_Y) \\
\mathbf{E}_3 &= (D_3, Y, (Y, H, X), f, (\mathcal{I}_1, \mathcal{I}_2, \mathcal{I}_3)) \\
\mathbf{E}_4 &= (D_4, H, in_H) \\
\mathbf{E}_5 &= (D_3, H, H, id, \mathcal{I}_2)
\end{aligned}
$$

where:

- the index mappings are:

$$\begin{aligned}
\mathcal{I}_1(i,j) &= (i, j-1) \\
\mathcal{I}_2(i,j) &= (i-1, j) \\
\mathcal{I}_3(i,j) &= (Mi - j + 1, 0)
\end{aligned}$$

- the applied functions are:

$$\begin{aligned}
f(a,b,c) &= a + bc \\
id(a) &= a \\
in_X(i,j) &= x(i) \\
in_Y(i,j) &= 0 \\
in_H(i,j) &= h(j-1)
\end{aligned}$$

- the domains are:

$$\begin{aligned}
D_1 &= \{(i,j) \mid i \geq -(N-1), j = 0\} \\
D_2 &= \{(i,j) \mid i \geq 1, j = 0\} \\
D_3 &= \{(i,j) \mid i \geq 1, 1 \leq j \leq N\} \\
D_4 &= \{(i,j) \mid i = 0, 1 \leq j \leq N\}
\end{aligned}$$

5.2.2 Analysis

The data dependencies of the system are the following:

$$\begin{aligned}
\mathcal{DD}_1 &= (D_3, Y, Y, \mathcal{I}_1) \\
\mathcal{DD}_2 &= (D_3, Y, H, \mathcal{I}_2) \\
\mathcal{DD}_3 &= (D_3, Y, X, \mathcal{I}_3) \\
\mathcal{DD}_4 &= (D_3, H, H, \mathcal{I}_2)
\end{aligned}$$

where the index mappings are integral and can be expressed, for instance, as:

$$\begin{aligned}
\mathcal{I}_1(i,j) &= (i,j) + (0, -1) \\
\mathcal{I}_2(i,j) &= (i,j) + (-1, 0) \\
\mathcal{I}_3(i,j) &= (i,j) + ((M-1)i + 1)(1,0) + j(-1, -1)
\end{aligned}$$

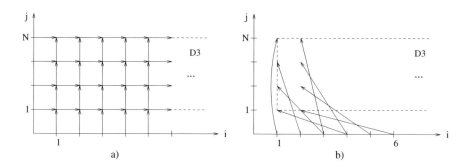

Fig. 5.17. Data dependence graph $(M = 3)$.

In particular, $\mathcal{I}_1$ and $\mathcal{I}_2$ are uniform. The form we have chosen for $\mathcal{I}_3$ guarantees that its coefficients define non-negative integer functions on D_3. Let $g_1(i, j) = (M - 1)i + 1$ and $g_2(i, j) = j$ be such coefficients.

The corresponding data dependence graph is sketched in Fig. 5.17, where, for clarity, we have separated the uniform data dependence vectors from those relative to $\mathcal{DD}_3$ (parts a) and b) of the figure, respectively). In the figure we have assumed $M = 3$.

A simple analysis of g_1 shows that the coefficient is not bounded on D_3. In fact, is value grows linearly in i. Therefore integral regularisation techniques are not applicable.

Although we have been treating $\mathcal{DD}_3$ as an integral data dependence, $\mathcal{DD}_3$ is in particular an affine data dependence. In fact, its index mapping can be written in matrix form as:

$$\mathcal{I}_3 \begin{pmatrix} i \\ j \end{pmatrix} = \begin{bmatrix} M & -1 \\ 0 & 0 \end{bmatrix} \begin{pmatrix} i \\ j \end{pmatrix} + \begin{pmatrix} 1 \\ 0 \end{pmatrix}$$

As the matrix $A = \begin{bmatrix} M & -1 \\ 0 & 0 \end{bmatrix}$ is rank deficient, a well known regularisation technique (known in the literature as *pipelining* – see [FoMo84, RaFu87, Raj89, QuVa89]) can be applied. According to this technique, a regularisation vector can be chosen as a non-null vector in $null(A) \cap lin(D_3)$ (see Appendix D for a definition of $null(A)$). D_3 is of full dimension, hence we can restrict ourselves to $null(A)$. The space $null(A)$ is spanned, for instance, by the vector $d = (1, M)$. As d is not a null vector, it can be chosen as a pipelining vector. The technique prescribes to partition the domain D_3 in two non-empty sub-sets defined as:

$$D_{3_1} \quad = \quad \{(i, j) \in D_3 \mid (i, j) + d \in D_3\}$$

$$D_{3_2} \;=\; \{(i,j) \in D_3 \mid (i,j) + d \notin D_3\}$$

Intuitively, D_{3_1} corresponds to computation points inside D_3, while D_{3_2} to points on (or around) some boundary of D_3. The result is obtained by pipelining the data among neighbour points in D_{3_1} according to the direction of d, while possible residual non-uniform data dependence vectors are confined to D_{3_2}.

Unfortunately, as $d = (1, M)$ depends on the problem parameter M, the applicability of the technique also depends on the values of M. In particular, for $M \geq N$, the subset D_{3_1} reduces to the empty set, and the technique is ineffective. Note that a different choice of pipelining vector would not solve the problem as any pipelining vector for the problem is an integer multiple of d.

5.2.3   Problem Revisited

Let us consider the original equation:

$$y(m) = \sum_{n=0}^{N-1} h(n)x(Mm - n)$$

For each m, $y(m)$ is a convolution of N terms, involving the filter responses $h(0), h(1), \ldots, h(N-1)$ and the input samples $x(Mm), x(Mm - 1), \ldots, x(Mm - N + 1)$. In other words, each $y(m)$ is just a convolution of N input samples. However, different from ordinary convolution, the selection of such input samples depends on the decimation rate M.

A straightforward formulation of the problem can be obtained by computing ordinary convolution and then decimating the outputs according to M. This can be obtained through the following equations, where y' computes all convolutions, while y is assigned the decimated samples only:

$$y'(m) \;=\; \sum_{n=0}^{N-1} h(n)x(m - n)$$

$$y(k) \;=\; y'(k)$$

for $m \geq 0$ and $k = 0, M, 2M, 3M, \ldots$.

The convolution problem is a very well known problem in regular array synthesis (see, e.g., [QuRo91, Meg92]). The decimation is easy to specify with the help of control signals: it simply amounts to locate a non-convex

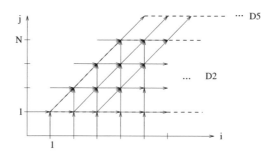

Fig. 5.18. Uniform data dependence graph.

sub-domain of a convex polyhedral domain. We have encountered this problem earlier in this chapter for cyclic reduction (see Section 5.1.1). In this
case, we may define a control variable C which assumes value equal to 1 on
those k of the domains which are integer multiples of M. A possible specification is the following system of equations (whose uniform data dependence
graph is illustrated in Fig. 5.18):

$$\text{(variable } X)$$
$$\mathbf{E}_1 \;=\; (D_1, X, in_X)$$
$$\mathbf{E}_2 \;=\; (D_2, X, X, id, \mathcal{I}_1)$$
$$\text{(variable } \bar{Y})$$
$$\mathbf{E}_3 \;=\; (D_3, \bar{Y}, in_{\bar{Y}})$$
$$\mathbf{E}_4 \;=\; (D_2, \bar{Y}, (\bar{Y}, H, X), f_1, (\mathcal{I}_2, \mathcal{I}_3, \mathcal{I}_1))$$
$$\text{(variable } H)$$
$$\mathbf{E}_5 \;=\; (D_4, H, in_H)$$
$$\mathbf{E}_6 \;=\; (D_2, H, H, id, \mathcal{I}_3)$$
$$\text{(control variable } C)$$
$$\mathbf{E}_7 \;=\; (D_5, C, in_C)$$
$$\text{(variable } Y)$$
$$\mathbf{E}_8 \;=\; (D_5, Y, (\bar{Y}, C), f_2, (\mathcal{I}_1, \mathcal{I}_0))$$

where:

- the index mappings are:

$$\mathcal{I}_0(i, j) \;=\; (i, j)$$

$$\begin{aligned}
\mathcal{I}_1(i,j) &= (i, j-1) \\
\mathcal{I}_2(i,j) &= (i-1, j-1) \\
\mathcal{I}_3(i,j) &= (i-1, j)
\end{aligned}$$

- the applied functions are:

$$\begin{aligned}
f_1(a,b,c) &= a + bc \\
f_2(a,b) &= \begin{cases} a & b=1 \\ \bot & b=0 \end{cases} \\
id(a) &= a \\
in_X(i,j) &= x(i-N) \\
in_{\bar{Y}}(i,j) &= 0 \\
in_H(i,j) &= h(N-j) \\
in_C(i,j) &= \begin{cases} 1 & (i-N-1) \bmod M = 0 \\ 0 & \text{otherwise} \end{cases}
\end{aligned}$$

- the domains are:

$$\begin{aligned}
D_1 &= \{(i,j) \mid i \geq 1, j = 0\} \\
D_2 &= \{(i,j) \mid i \geq j, 1 \leq j \leq N\} \\
D_3 &= \{(i,j) \mid i \geq 0, j = 0\} \\
D_4 &= \{(i,j) \mid i = j-1, 1 \leq j \leq N\} \\
D_5 &= \{(i,j) \mid i \geq j, j = N+1\}
\end{aligned}$$

5.2.4   Space-Time Mapping

All data dependencies are uniform and the corresponding dependence cone is

$$\Theta^* = cone(\{(0,1), (1,1), (1,0)\}),$$

with extremal rays $(0,1)$ and $(1,0)$. The dependence cone is illustrated in Fig. 5.19 a).

As the dependence cone is pointed, a valid affine timing function is determined by any $\lambda = (\lambda_1, \lambda_2) \in \mathbf{Z}^2$ such that the following system of inequalities is satisfied:

$$\begin{aligned}
\lambda_1 &> 0 \\
\lambda_2 &> 0.
\end{aligned}$$

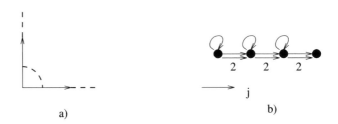

a)   b)

Fig. 5.19. a) Dependence cone; b) Signal flow graph.

d	$\lambda \cdot d$	$a(d)$
$(1,0)$	1	0
$(1,1)$	2	1
$(0,1)$	1	1

Table 5.9. Transformation of data dependence vectors under the space-time mapping $[t, a]$.

The vector $\lambda = (1,1)$ is a possible choice. A non-negative timing function determined by λ is the function $t(i,j) = i + j$.

A compatible linear allocation function is a projection according to any non-null vector $u = (u_1, u_2) \in \mathbf{Z}^2$, such that $\lambda \cdot u \neq 0$, i.e., $u \in \mathbf{Z}^2$ such that $u_1 + u_2 \neq 0$. However, a finite array design can be obtained only if the projection is done in the same direction of the ray of the domain D_2, that is for a projection vector $u = (1,0)$, which corresponds to the allocation function $a(i,j) = j$.

The transformation of the dependence vectors of the system under the space-time mapping $[t, a]$ is summarised in Table 5.9. The resulting signal flow graph is illustrated in Fig. 5.19 b). For simplicity arcs with a unit delay are not labelled. A description of the corresponding basic processing element and its operations is given in Fig. 5.20.

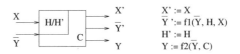

Fig. 5.20. Processing element.

5.3   Knapsack Problem

In this section we consider the knapsack problem as an example of dynamic problem. The knapsack problem can be formulated either as a linear optimisation problem or as a dynamic programming problem (see, e.g., [Hu82]). The latter formulation is based on recursive functions, and is the formulation we have chosen in this section.

The problem can be described as follows. Let us consider a knapsack of finite capacity c, together with objects of n different types, each type i, for $i = 1, \ldots, n$, characterised by a weight w_i and a value v_i. We assume that both the capacity of the knapsack and the weights of the objects are integers, and that, for all i, $0 < w_i \leq c$. We also assume that there exists an unbounded number of objects of each type. The knapsack problem consists of determining the optimal (i.e., the most valuable) selection of the given objects, which can be carried in the knapsack without exceeding its capacity.

A possible dynamic programming formulation of the algorithm (based on [Hu82]) is the following. Let $f_i(j)$ represent the maximum value which can be carried in a knapsack of capacity j by selecting objects of types from 1 to i. $f_i(j)$ may be defined as follows:

$$
\begin{array}{lll}
i = 1, \ldots, n, & j = 1, \ldots, c & f_i(j) = max(f_{i-1}(j), f_i(j - w_i) + v_i) \\
i = 0, & j = 0, \ldots, c, & f_i(j) = 0 \\
i = 0, \ldots n, & j = 0 & f_i(j) = 0 \\
i = 1, \ldots n, & j < 0 & f_i(j) = -\infty
\end{array}
$$

The boundary conditions express that $f_i(j)$ is 0 if either the capacity of the knapsack is equal to 0 or there are no objects which can be selected; and $f_i(j)$ is $-\infty$ if the capacity of the knapsack is negative, where the symbol ∞ indicates the largest integer representable in a machine. The solution to the knapsack problem is the value $f_n(c)$.

Note that these equations only compute the optimal value of the objects which can be carried in the knapsack. In the literature this is known as the forward phase of the algorithm [Hu82]. The corresponding combination of objects can be determined subsequently as a backward substitution process on the sub-optimal values of f. This is known as the backward phase of the algorithm. This phase is essentially sequential and will not be considered here. (An efficient algorithm for the backward phase is given in [Hu82].)

5.3.1 Specification

In order to specify the algorithm as a system of recurrence equations, we make the following observations:

- two indices, i and j, are sufficient to the expression of the recurrences. Hence, we adopt $\mathbf{Z}^2$ as the computation space and associate i to the types of the objects and j to the capacity of the knapsack;

- a variable F, corresponding to f, is needed to compute the optimal solution, together with two variables, W and V respectively, for the weights and values of the objects;

- the computation domain of F is the rectangular region determined by the points (i, j), such that $i = 1, \ldots, n$ and $j = 1, \ldots, c$. Variables W and V are initialised on the boundary of this domain and their values are subsequently pipelined through the region.

A possible system of equations is the following:

$$\text{(variable } F)$$
$$\mathbf{E}_1 = (D_1, F, in_F^1)$$
$$\mathbf{E}_2 = (D_2, F, in_F^1)$$
$$\mathbf{E}_3 = (D_3, F, in_F^2)$$
$$\mathbf{E}_4 = (D_4, F, (F, F, V), f, (\mathcal{I}_1, \mathcal{I}_2, \mathcal{I}_3))$$
$$\text{(variable } V)$$
$$\mathbf{E}_5 = (D_2, V, in_V)$$
$$\mathbf{E}_6 = (D_4, V, V, id, \mathcal{I}_3)$$
$$\text{(variable } W)$$
$$\mathbf{E}_7 = (D_2, W, in_W)$$
$$\mathbf{E}_8 = (D_4, W, W, id, \mathcal{I}_3)$$

whit:

- index mappings:

$$\mathcal{I}_1(i, j) = (i - 1, j)$$
$$\mathcal{I}_2(i, j) = (i, j - W(i, j))$$
$$\mathcal{I}_3(i, j) = (i, j - 1)$$

- applied functions:

$$
\begin{aligned}
in_F^1(i,j) &= 0 \\
in_F^2(i,j) &= -\infty \\
in_V(i,j) &= v_i \\
in_W(i,j) &= w_i \\
f(a,b,c) &= max(a, b+c)
\end{aligned}
$$

where w_i is an integer, with $0 < w_i \le c$, and $v_i \ge 0$;

- domains:

$$
\begin{aligned}
D_1 &= \{(i,j) \mid i = 0, 1 \le j \le c\} \\
D_2 &= \{(i,j) \mid 1 \le i \le n, j = 0\} \\
D_3 &= \{(i,j) \mid 1 \le i \le n, j < 0\} \\
D_4 &= \{(i,j) \mid 1 \le i \le n, 1 \le j \le c\}.
\end{aligned}
$$

5.3.2 Analysis

The data dependencies of the specification are the following:

$$
\begin{aligned}
\mathcal{DD}_1 &= (D_4, F, F, \mathcal{I}_1) \\
\mathcal{DD}_2 &= (D_4, F, F, \mathcal{I}_2) \\
\mathcal{DD}_3 &= (D_4, F, V, \mathcal{I}_3) \\
\mathcal{DD}_4 &= (D_4, V, V, \mathcal{I}_3) \\
\mathcal{DD}_5 &= (D_4, W, W, \mathcal{I}_3)
\end{aligned}
$$

with index mappings:

$$
\begin{aligned}
\mathcal{I}_1(i,j) &= (i-1,j) \\
\mathcal{I}_2(i,j) &= (i, j - W(i,j)) \\
\mathcal{I}_3(i,j) &= (i, j-1)
\end{aligned}
$$

Data dependencies $\mathcal{DD}_1, \mathcal{DD}_3, \mathcal{DD}_4$ and $\mathcal{DD}_5$ are static and uniform, while data dependence $\mathcal{DD}_2$ is dynamic and atomic finitely generated. An explicit form for the index mapping is:

$$
\begin{aligned}
\mathcal{I}_1(i,j) &= (i,j) + (-1, 0) \\
\mathcal{I}_2(i,j) &= (i,j) + W(i,j)(0, -1) \\
\mathcal{I}_3(i,j) &= (i,j) + (0, -1)
\end{aligned}
$$

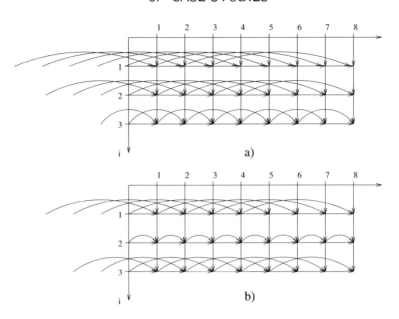

Fig. 5.21. Two instances of the dynamic dependence graph of the knapsack problem, for $c = 8$, $n = 3$, and weight distributions, respectively: a) $w_1 = 5, w_2 = 4, w_3 = 2$; b) $w'_1 = 4, w'_2 = 1, w'_3 = 3$.

where $0 < W(i,j) \le c$, for all (i,j). As $\mathcal{DD}_2$ is dynamic, for each configuration of weights in input we obtain a different data dependence graph. For $c = 8$ and $n = 3$, two instances of the dynamic data dependence graph are given in Fig. 5.21 a) and b), corresponding to the two distributions of weights $w_1 = 5, w_2 = 4, w_3 = 2$ and $w'_1 = 4, w'_2 = 1, w'_3 = 3$, respectively. Note that only the sub-graph relative to $\mathcal{DD}_2$ varies with the inputs.

By considering the generators of all index mappings, we obtain the embedding dependence cone $C = cone(\{(1,0),(0,1)\}$, illustrated in Fig. 5.22 a). Note that C is pointed and, because of the condition $0 < W(i,j) \le c$, it contains the dependence cone of the specification for all inputs (see also the discussion in Section 4.3.3).

The separability of F and W, which is necessary for uniformisation, can be verified on the extended dependence graph of the system. This graph is illustrated in Fig. 5.22 b), and is defined as the graph $\mathcal{EDG} = (\mathcal{N}, \mathcal{A})$, where $\mathcal{N} = \{F, V, W\}$ and $\mathcal{A} = \{(F, F), (F, V), (F, W), (V, V), (W, W)\}$. From the graph we can deduce that variables F and W are separable in $\mathcal{EDG}$.

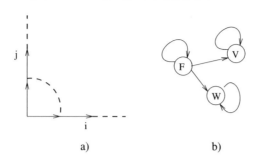

Fig. 5.22. a) Cone C; b) Extended dependence graph.

5.3.3   Regularisation

We want to make the data dependence $\mathcal{D}\mathcal{D}_2$ uniform, by applying parametric uniformisation according to Proposition 4.4.11. Let $p \in \mathbf{N}^+$. We note that:

- for all (i, j), $0 < W(i, j) \leq c$, then $\bar{W} = \lfloor c/(p+1) \rfloor$.

- $lin(D) = \mathbf{Z}^2$ and $d = (0, -1) \in lin(D)$, hence we need to reindex the system in $\mathbf{Z}^3$. Let k indicate the index of the added axis. We choose the new system of axes characterised by the indices i, j, k in this order.

- π can be chosen as the vector $\pi = (0, 0, 1)$, where π is in $D_4^{\perp}$ and D_4 is contained in the hyperplane $[\pi : 0]$. Then $\hat{d} = d + \pi = (0, -1, 1)$ and $\check{d} = d - \pi = (0, -1, -1)$.

- tr is the translation defined as $tr(i, j, k) = (i, j, k) + (0, -\bar{W}, \bar{W}) = (i, j - \bar{W}, k + \bar{W})$, and ren renames W as W^{tr}.

- the system of equations defining W is $Def\mathbf{S}_W = \{\mathbf{E}_7, \mathbf{E}_8\}$, and its image under tr and ren is $(Def\mathbf{S}_W)^{tr,ren}$ defined as:

$$
\begin{aligned}
\mathbf{E}_7' &= (D_2^{tr}, W^{tr}, in_{W^{tr}}) \\
\mathbf{E}_8' &= (D_4^{tr}, W^{tr}, W^{tr}, id, \mathcal{I}_3^{tr})
\end{aligned}
$$

with domains:

$$
\begin{aligned}
D_2^{tr} &= \{(i, j, k) \mid 1 \leq i \leq n, j = -\bar{W}, k = \bar{W}\} \\
D_4^{tr} &= \{(i, j, k) \mid 1 \leq i \leq n, 1 - \bar{W} \leq j \leq c - \bar{W}, k = \bar{W}\}
\end{aligned}
$$

and index mapping:

$$
\mathcal{I}_3^{tr}(i, j, k) = t \circ \mathcal{I}_3 \circ t^{-1}(i, j, k) = (i, j, k) + (0, -1, 0).
$$

The parametric uniformisation of $\mathcal{DD}_2$ produces the system of equations:

$\quad\quad\quad$ (variable F)

$$
\begin{aligned}
\mathbf{E}_1 &= (D_1, F, in_F^1) \\
\mathbf{E}_2 &= (D_2, F, in_F^1) \; \cdot \\
\mathbf{E}_3 &= (D_3, F, in_F^2) \\
\mathbf{E}_4 &= (D_4, F, (F, R^1, V), f, (\mathcal{I}_1, \mathcal{I}_2, \mathcal{I}_3))
\end{aligned}
$$

$\quad\quad\quad$ (variable V)

$$
\begin{aligned}
\mathbf{E}_5 &= (D_2, V, in_V) \\
\mathbf{E}_6 &= (D_4, V, V, id, \mathcal{I}_3)
\end{aligned}
$$

$\quad\quad\quad$ (variable W^{tr})

$$
\begin{aligned}
\mathbf{E}_7 &= (D_2^{tr}, W^{tr}, in_{W^{tr}}) \\
\mathbf{E}_8 &= (D_4^{tr}, W^{tr}, W^{tr}, id, \mathcal{I}_3^{tr})
\end{aligned}
$$

$\quad\quad\quad$ (routing variables R^1, R^2 and R^3)

$$
\begin{aligned}
\mathbf{E}_9 &= (D_{4_1}, R^1, (\alpha, \beta, \gamma, R^1, R^2, \ldots, R^2), f', (\mathcal{I}_2, \mathcal{I}_2, \mathcal{I}_2, \mathcal{I}_4, \mathcal{I}_{5,0}, \ldots, \mathcal{I}_{5,p})) \\
\mathbf{E}_{10} &= (D_{4_{2,1}}, R^2, R^3, id, \mathcal{I}_6) \\
\mathbf{E}_{11} &= (D_{4_{2,2}}, R^2, F, id, \mathcal{I}_2) \\
\mathbf{E}_{12} &= (D_{4_{2,3}}, R^3, R^2, id, \mathcal{I}_7)
\end{aligned}
$$

$\quad\quad\quad$ (control variables α, β and γ)

$$
\begin{aligned}
\mathbf{E}_{13} &= (D_{4_{1,1}}, \alpha, \alpha, id, \mathcal{I}_4) \\
\mathbf{E}_{14} &= (D_{4_{1,2}}, \alpha, W^{tr}, (p+1)\_floor, \mathcal{I}_2) \\
\mathbf{E}_{15} &= (D_{4_{1,1}}, \beta, \beta, id, \mathcal{I}_4) \\
\mathbf{E}_{16} &= (D_{4_{1,2}}, \beta, W^{tr}, mod_{p+1}, \mathcal{I}_2) \\
\mathbf{E}_{17} &= (D_{4_{1,1}}, \gamma, \gamma, dec, \mathcal{I}_4) \\
\mathbf{E}_{18} &= (D_{4_{1,2}}, \gamma, in_\gamma)
\end{aligned}
$$

with:

$\quad$ - index mappings:

$$
\begin{aligned}
\mathcal{I}_1(i, j, k) &= (i, j, k) + (-1, 0, 0) \\
\mathcal{I}_2(i, j, k) &= (i, j, k) \\
\mathcal{I}_3(i, j, k) &= (i, j, k) + (0, -1, 0) \\
\mathcal{I}_3^{tr}(i, j, k) &= (i, j, k) + (0, -1, 0)
\end{aligned}
$$

$$\begin{aligned}
\mathcal{I}_4(i,j,k) &= (i,j,k) + (0,-1,1) \\
\mathcal{I}_{5,0}(i,j,k) &= (i,j,k) \\
\mathcal{I}_{5,1}(i,j,k) &= i,j,k + (0,-1,0)
\end{aligned}$$

$$\cdots$$

$$\begin{aligned}
\mathcal{I}_{5,p}(i,j,k) &= (i,j,k) + (0,-p,0) \\
\mathcal{I}_6(i,j,k) &= (i,j,k) + (0,-1,-1) \\
\mathcal{I}_7(i,j,k) &= (i,j,k) + (0,1-p,0)
\end{aligned}$$

- applied functions:

$$\begin{aligned}
in_F^1(i,j,k) &= 0 \\
in_F^2(i,j,k) &= -\infty \\
in_V(i,j,k) &= v_i \\
in_W^{tr}(i,j,k) &= w_i \\
in_\gamma(i,j,k) &= \bar{W} \\
(p+1)\_floor(a) &= \lfloor a/(p+1) \rfloor \\
mod_{p+1}(a) &= a \bmod (p+1) \\
id(a) &= a \\
dec(a) &= a-1 \\
f(a,b,c) &= max(a,b+c) \\
f'(\alpha,\beta,\gamma,a_1,a_{2,0},\ldots,a_{2,p}) &= \begin{cases} a_1 & \alpha \neq \gamma \\ a_{2,0} & \alpha = \gamma, \beta = 0 \\ \cdots & \\ a_{2,p} & \alpha = \gamma, \beta = p \end{cases}
\end{aligned}$$

- domains:

$$\begin{aligned}
D_1 &= \{(i,j,k) \mid i=0, 1 \le j \le c, k=0\} \\
D_2 &= \{(i,j,k) \mid 1 \le i \le n, j=0, k=0\} \\
D_3 &= \{(i,j,k) \mid 1 \le i \le n, j<0, k=0\} \\
D_4 &= \{(i,j,k) \mid 1 \le i \le n, 1 \le j \le c, k=0\} \\
D_{4_1} &= \{(i,j,k) \mid 1 \le i \le n, 0 \le k \le \bar{W}, 1-k \le j \le c-k\} \\
D_{4_{1,1}} &= \{(i,j,k) \in D_{4_1} \mid k < \bar{W}\} \\
D_{4_{1,2}} &= \{(i,j,k) \in D_{4_1} \mid k = \bar{W}\}
\end{aligned}$$

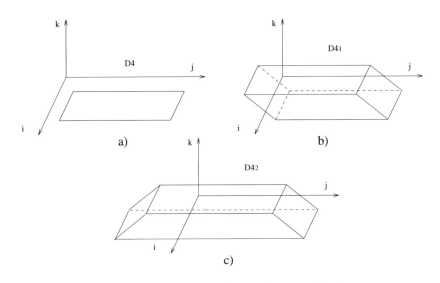

Fig. 5.23. Domains: a) D_4; b) D_{4_1}; c) D_{4_2}.

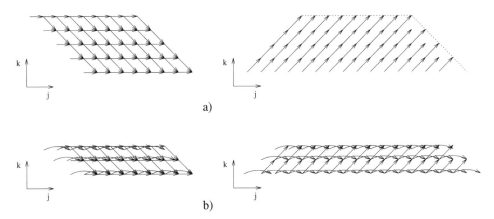

Fig. 5.24. Sections of the routing domains D_{4_1} and D_{4_2} for values of the parameter: a) $p = 1$; b) $p = 2$.

$$
\begin{aligned}
D_{4_2} &= \{(i,j,k) \mid 1 \le i \le n, 0 \le k \le \bar{W}, k - 2\bar{W} \le j \le c - k\} \\
D_{4_2,1} &= \{(i,j,k) \in D_{4_2} \mid k > 0\} \\
D_{4_2,2} &= \{(i,j,k) \in D_{4_2} \mid k = 0\} \\
D_{4_2,3} &= \{(i,j,k) \in D_{4_2} \mid k < \bar{W}\}.
\end{aligned}
$$

Pair	*Unifor.*	*DD Vector*
F, F	$-$	$(1, 0, 0)$
F, F	F, R^1	$(0, 0, 0)$
	R^1, R^1	$(0, 1, -1)$
	R^1, R^2	$(0, 0, 0)$
	R^1, R^2	$(0, 1, 0)$
	$\cdots$	
	R^1, R^2	$(0, p, 0)$
	R^2, R^3	$(0, 1, 1)$
	R^3, R^2	$(0, p-1, 0)$
	R^2, F	$(0, 0, 0)$
F, V	$-$	$(0, 1, 0)$
V, V	$-$	$(0, 1, 0)$
W^{tr}, W^{tr}	$-$	$(0, 1, 0)$
α, α	$-$	$(0, 1, -1)$
β, β	$-$	$(0, 1, -1)$
γ, γ	$-$	$(0, 1, -1)$

Table 5.10. Summary of the data dependencies.

Domains D_4, D_{4_1} and D_{4_2} are sketched in Fig. 5.23, a), b) and c), respectively, while Fig. 5.24 a) and b) illustrate sections of the data dependence graphs in D_{4_1} and D_{4_2} for values of the parameter $p = 1, 2$, respectively (in the figure, we assume $c = 8$). Only the routing variables R^1, R^2 and R^3 are considered.

5.3.4 Space-Time Mapping

After regularisation, all data dependencies are uniform. They are summarised in Table 5.10 (with similar conventions to those in Section 5.1.6). The corresponding dependence cone is

$$\Theta^* = cone(\{(0, 1, 0), (0, 2, 0), \ldots, (0, p, 0), (0, 1, -1), (0, 1, 1), (1, 0, 0)\}),$$

with extremal rays $(0, 1, -1)$, $(0, 1, 1)$ and $(1, 0, 0)$. The dependence cone, illustrated in Fig. 5.25, is pointed.

A valid affine timing function is determined by any $\lambda = (\lambda_1, \lambda_2, \lambda_2) \in \mathbf{Z}^3$ such that the following system of inequalities is satisfied:

$$\lambda_2 - \lambda_3 > 0$$

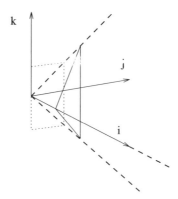

Fig. 5.25. Dependence cone after uniformisation.

d	$\lambda \cdot d$	$a(d)$	$a'(d)$	$a''(d)$
$(1,0,0)$	1	$(1,0)$	$(1,0)$	$(0,0)$
$(0,1,0)$	1	$(0,0)$	$(-1,0)$	$(1,0)$
$(0,2,0)$	2	$(0,0)$	$(-2,0)$	$(2,0)$
$\dots$	$\dots$	$\dots$	$\dots$	$\dots$
$(0,p,0)$	p	$(0,0)$	$(-p,0)$	$(p,0)$
$(0,1,-1)$	1	$(0,-1)$	$(-1,-1)$	$(1,-1)$
$(0,1,1)$	1	$(0,1)$	$(-1,1)$	$(1,1)$

Table 5.11. Transformation of data dependence vectors under the space-time mappings $[t, a]$, $[t, a']$, and $[t, a'']$.

$$\lambda_2 + \lambda_3 > 0$$
$$\lambda_1 > 0$$

The vector $\lambda = (1,1,0)$ is a possible choice, corresponding to the timing function $t(i, j, k) = i+j$. With this timing function, the algorithm is executed in $O(c + 2\lfloor c/2 \rfloor + n)$ steps.

A compatible linear allocation function is a projection according to any non-null vector $u = (u_1, u_2, u_3) \in \mathbf{Z}^3$, such that $\lambda \cdot u \neq 0$, i.e., $u \in \mathbf{Z}^3$ such that $u_1 + u_2 \neq 0$. A possible projection vector is $u = (0,1,0)$, corresponding to the allocation function $a(i, j, k) = (i, k)$. The transformation of the dependence vectors of the system under the space-time mapping $[t, a]$ is summarised in Table 5.11.

The resulting signal flow graph is illustrated in Fig. 5.26 a), where the

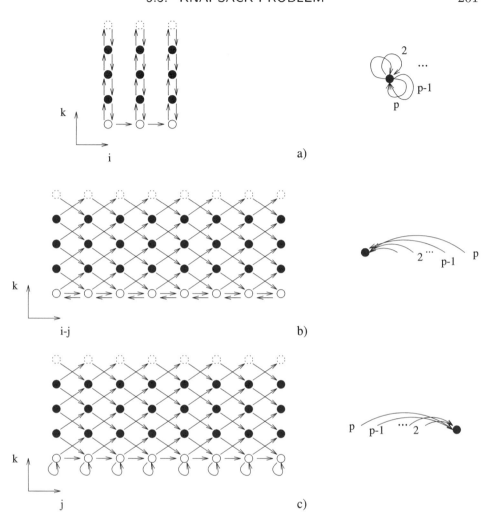

Fig. 5.26. Signal flow graphs for: a) $[t, a]$; b) $[t, a']$; c) $[t, a'']$.

left-hand side illustrates the nodes and the non-parametric communication channels, while the right-hand side details the parametric communication channels at each node. For simplicity arcs with a unit delay are not labelled. Note that there are three types of nodes in the graph (illustrated as white, black and dotted nodes).

Under $[t, a]$, the p parametric communication links may be implemented as p locations of the local RAM at each processing element of the array. The three types of cells are described in Fig. 5.27 a).

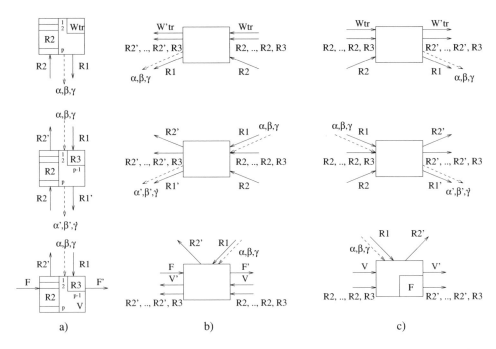

Fig. 5.27. Processing elements corresponding to: a) $[t, a]$; b) $[t, a']$; c) $[t, a'']$.

Different compatible placements are determined, for instance, by the projection vectors $u' = (1, 1, 0)$ and $u'' = (1, 0, 0)$, corresponding to the allocation functions $a'(i, j, k) = (i - j, k)$ and $a''(i, j, k) = (j, k)$, respectively. The transformation of the dependence vectors under $[t, a']$ and $[t, a'']$ are also given in Table 5.11. The resulting signal flow graphs are illustrated in Fig. 5.26 b) and c), while the corresponding processing elements are sketched in Fig. 5.27 b) and c), respectively.

5.4   Gaussian Elimination with Partial Pivoting

Gaussian elimination reduces an $n \times n$ matrix $A = [a_{i,j}]$ to a triangular form in $n - 1$ iterations. At each iteration the elements, under the main diagonal, of one of the columns of the matrix are nullified by a series of elementary row operations[1]. The columns are processed from left to right. At each iteration

---

[1]Elementary row operations correspond to changes of basis of a vector space [Ner63]. They include: swapping rows, multiplying a row by a non-zero scalar; and subtracting from one row a multiple of another row.

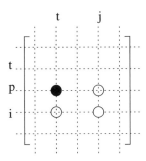

Fig. 5.28. Pivot operation: the black circle indicates the pivot; the white circle indicates the entry to be updated; and the other circles indicate the corresponding row and column coefficients.

t, for $1 \leq t \leq n - 1$, a new matrix is generated as follows. The element $a_{t,t}$ of the current matrix is selected and used to update each $a_{i,j}$, with $i > t$ and $j \geq t$, by applying the operation $a_{i,j} - a_{i,t} \cdot a_{t,j}/a_{t,t}$.

When Gaussian elimination is combined with partial pivoting, at iteration t, an element of the subcolumn $[a_{t,t}, a_{t+1,t}, \ldots, a_{n,t}]$ with maximum modulo is chosen as the *pivot* at that iteration and used to update the current matrix. Let $a_{p,t}$ denote the pivot at iteration t. The new matrix is updated by applying the operation $a_{i,j} - a_{i,t} \cdot a_{p,j}/a_{p,t}$, for all i, j such that $i \geq t, i \neq p$ and $j \geq t$. (By generalising, Gaussian elimination without pivoting may be seen as a form of pivoting in which at each iteration t, $a_{p,t} = a_{t,t}$.)

A way of exploiting parallelism in Gaussian elimination with pivoting is to break the computation of the operation which updates the elements of the matrix into sub-computations, which can be executed in a distributed fashion. Let us consider the operation $a_{i,j} - a_{i,t} \cdot a_{p,j}/a_{p,t}$. The quotient $a_{i,t}/a_{p,t}$ is used to update all the elements of the ith row of the matrix, while $a_{p,j}$ is applied to all the elements of the jth column. This is illustrated in Fig. 5.28. These values can be treated as coefficients, which can be propagated among computation points and used for the parallel updating of the matrix. In the following we will call them row and column coefficients, respectively.

In Gaussian elimination without pivoting, at each iteration a particular element of the matrix is selected according to *its position* in the matrix (namely, $a_{t,t}$, at iteration t). When pivoting is present, the pivot is selected according to *its absolute value* as an entry of the current matrix, and this value is known at run-time only, i.e., when the algorithm is executed on

actual data. It is this characteristics which makes Gaussian elimination
with pivoting a dynamic problem.

5.4.1 Dynamic Formulation

In order to specify Gaussian elimination with pivoting as a dynamic problem,
let us consider the basic operation

$$a_{i,j} - a_{i,t} \cdot a_{p,j}/a_{p,t}$$

more closely. The index p indicating the row of the pivot is actually a
function of the iteration index t, and should be indicated more precisely as
$p(t)$. Therefore, both a and p can be seen as variables to be computed by the
algorithm. Then at each iteration t, for $t = 1, \ldots, n$, the algorithm computes
the values:

$$
\begin{aligned}
p(t) &= \odot\{a_{i,t}^{t-1} \mid t \le i \le n\} \\
a_{i,j}^{t} &= a_{i,j}^{t-1} - a_{i,t}^{t-1} * a_{p(t),j}^{t-1}/a_{p(t),t}^{t-1}
\end{aligned}
$$

for all i, j such that $i \ge t$ and $i \ne p(t)$. The operation $\odot$ compares the entries
of the tth column of the current matrix to find the pivot, and once found it,
returns the corresponding row index. The initial values are $a_{i,j}^{0} = a_{i,j}$ for all
i, j.

If we had to specify the problem more formally as a system of dynamic
recurrence equations we would soon realise that the techniques we have developed are not powerful enough to deal with this problem, the reason being
that the computations of $p(t)$ and the updating of the entries $a_{i,j}^{t}$ of the
matrix are interleaved at each iteration. This fact introduces a mutual dependence between the computations of the pivot and the new entries of the
current matrix so that the variables corresponding to p and a are not separable.

5.4.2 Static Formulation

Regular array designs for Gaussian elimination with partial pivoting have
been proposed in the literature, which are based on a static formulation of
the problem [BaEl88, Meg90, ElBa90]. This is obtained if, at each iteration,
the elements of the current matrix are rearranged before the new entries are
computed. In this way the pivot always assumes a predetermined position.

Each iteration of the new algorithm includes the following operations: finding the pivot and rearranging the elements of the matrix; determining and propagating the row and column coefficients; and computing the new entries of the matrix. We note that:

- at each iteration of the algorithm the same sequence of operations is carried out on the current matrix: hence the specification of the algorithm can be simplified by concentrating on the single iterations (otherwise we would need to consider a specification in 4 dimensions);

- the size of the current matrix decreases at each iteration: hence the specification of the iterations can be induced from the specification of the first iteration by a restriction to appropriate submatrices;

- the output matrix of each iteration represents the input matrix of the following iteration: hence a strong sequentiality constraint exists between consecutive iterations, and the overall design of the algorithm can be obtained as a sequential composition of the specifications of the single iterations. Indeed pipelining techniques should be used to overlap some of the operations of consecutive iterations.

Preliminary Specification

At each iteration of the algorithm, the pivot is chosen as the element with maximum modulo in a set of entries of the matrix. In this section we illustrate a technique for the selection of the pivot, which is used in the specifications of the following sections. The technique is a sorting of the elements according to their absolute value.

Let us consider n numbers, $a_1, \ldots, a_n$. Sorting the n numbers $a_1, \ldots, a_n$ according to their absolute value can be realised as follows. We introduce two variables V, H, with V initialised with the n given numbers (plus $n-1$ zero entries) and H with all its $n-1$ entries equal to ∞. The computations of V and H in a two dimensional space are illustrated by the data dependence graph in Fig. 5.29 (in the figure we have assumed $n = 4$). At each point (i, j) the pair of values $V(i-1, j)$ and $H(i, j-1)$ are compared. The element with greater absolute value is stored in $V(i, j)$, while the other element in $H(i, j)$. The recurrences are:

$$\text{(variable } V)$$
$$\mathbf{E}_1 \quad = \quad (D_1, V, in_{V1})$$

$$\begin{aligned}
\mathbf{E}_2 &= (D_2, V, in_{V2}) \\
\mathbf{E}_3 &= (D_4, V, (V, H), |max|, (\mathcal{I}_1, \mathcal{I}_2)) \\
&\quad (\text{variable } H) \\
\mathbf{E}_4 &= (D_3, H, in_H) \\
\mathbf{E}_5 &= (D_4, H, (V, H), |min|, (\mathcal{I}_1, \mathcal{I}_2))
\end{aligned}$$

where:

- the index mappings are:

$$\begin{aligned}
\mathcal{I}_1(i, j) &= (i - 1, j) \\
\mathcal{I}_2(i, j) &= (i, j - 1)
\end{aligned}$$

- the applied functions are:

$$\begin{aligned}
in_{V1}(i, j) &= a_j \\
in_{V2}(i, j) &= 0 \\
in_H(i, j) &= \infty \\
|max|(a, b) &= \begin{cases} a & a \geq b \\ b & \text{otherwise} \end{cases} \\
|min|(a, b) &= \begin{cases} b & a \geq b \\ a & \text{otherwise} \end{cases}
\end{aligned}$$

- the domains are:

$$\begin{aligned}
D_1 &= \{(i, j) \mid i = 0, 1 \leq j \leq n\} \\
D_2 &= \{(i, j) \mid 0 \leq i \leq n - 2, j = i + n + 1\} \\
D_3 &= \{(i, j) \mid 1 \leq i \leq n - 1, j = i - 1\} \\
D_4 &= \{(i, j) \mid 1 \leq i \leq n - 1, i \leq j \leq i + n\}.
\end{aligned}$$

Specification

At each iteration t, the subcolumn $[a_{t,t}, a_{t+1,t}, \ldots, a_{n,t}]$ of the matrix is scanned to identify the pivot. Every time two elements of the column are swapped, their corresponding rows are also exchanged. This provides the

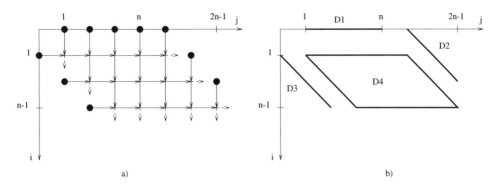

Fig. 5.29. Sorting: a) data dependence graph; b) domains.

necessary rearrangement of the elements of the matrix before updating its entries. Once all the rows of the matrix have been rearranged, the row and column coefficients are determined and the new matrix computed.

Let us consider the first iteration of the algorithm, i.e., $t = 1$. The iteration can be represented in a 3-dimensional space, in which we assume that A is input on the (k, j)-plane (the index k corresponds to columns of A). For $k = 1$, the elements of the first column of A are processed, looking for the pivot, as explained in the previous section. When the values of $H(i, j-1, 1)$ and $V(i-1, j, 1)$ are swapped, a control signal (variable C below) is issued and propagated upwards (direction of $(0, 0, 1)$). In this way, for $k > 1$, the corresponding exchange of $H(i, j-1, k)$ and $V(i-1, j, k)$ is performed. The first phase of the iteration can be specified as follows:

$$
\begin{aligned}
& \qquad \text{(variable } V) \\
\mathbf{E}_1 &= (D_1, V, in_{V1}) \\
\mathbf{E}_2 &= (D_2, V, in_{V2}) \\
\mathbf{E}_3 &= (D_{4,1}, V, (V, H), |max|, (\mathcal{I}_1, \mathcal{I}_2)) \\
\mathbf{E}_4 &= (D_{4,2}, V, (V, H, C), swap_V, (\mathcal{I}_1, \mathcal{I}_2, \mathcal{I}_3)) \\
& \qquad \text{(variable } H) \\
\mathbf{E}_5 &= (D_3, H, in_H) \\
\mathbf{E}_6 &= (D_{4,1}, H, (V, H), |min|, (\mathcal{I}_1, \mathcal{I}_2)) \\
\mathbf{E}_7 &= (D_{4,2}, H, (V, H, C), swap_H, (\mathcal{I}_1, \mathcal{I}_2, \mathcal{I}_3)) \\
& \qquad \text{(control variable } C) \\
\mathbf{E}_8 &= (D_{4,1}, C, (V, H), |cmp|, (\mathcal{I}_1, \mathcal{I}_2))
\end{aligned}
$$

$$\mathbf{E}_9 \;=\; (D_{4,2}, C, C, id, \mathcal{I}_3)$$

where:

- the index mappings are:

$$
\begin{aligned}
\mathcal{I}_1(i,j,k) &= (i-1,j,k) \\
\mathcal{I}_2(i,j,k) &= (i,j-1,k) \\
\mathcal{I}_3(i,j,k) &= (i,j,k-1)
\end{aligned}
$$

- the applied functions are:

$$
\begin{aligned}
in_{V1}(i,j,k) &= a_{j,k} \\
in_{V2}(i,j,k) &= 0 \\
in_H(i,j,k) &= \infty \\
id(a) &= a \\
swap_V(a,b) &= b \\
swap_H(a,b) &= a \\
|max|(a,b) &= \begin{cases} a & a \geq b \\ b & \text{otherwise} \end{cases} \\
|min|(a,b) &= \begin{cases} b & a \geq b \\ a & \text{otherwise} \end{cases} \\
|cmp|(a,b) &= \begin{cases} 0 & a \geq b \\ 1 & \text{otherwise} \end{cases}
\end{aligned}
$$

- the domains are:

$$
\begin{aligned}
D_1 &= \{(i,j,k) \mid i=0, 1 \leq j \leq n, 1 \leq k \leq n\} \\
D_2 &= \{(i,j,k) \mid 0 \leq i \leq n-2, j = i+n+1, 1 \leq k \leq n\} \\
D_3 &= \{(i,j,k) \mid 1 \leq i \leq n-1, j = i-1, 1 \leq k \leq n\} \\
D_{4,1} &= \{(i,j,k) \mid 1 \leq i \leq n-1, i \leq j \leq i+n, k=1\} \\
D_{4,2} &= \{(i,j,k) \mid 1 \leq i \leq n-1, i \leq j \leq i+n, 2 \leq k \leq n\}.
\end{aligned}
$$

$|cmp|$ compares the absolute values of its arguments and sets the signal C to 1 if an exchange is necessary (0 otherwise). $swap_H$ and $swap_V$ swap the values of H and V if the signal C is set to 1 (they pipeline H and

V, respectively, otherwise). At the end of this phase, $V(n-1,j,1)$, for $n \leq j \leq 2n-1$, contain the elements of the pivot column, sorted in decreasing absolute value, and $V(n-1,n,1)$ is the pivot. These elements are used to compute the row coefficients, which are subsequently pipelined upwards (direction of $(0,0,1)$). The pivot row is contained in $V(n-1,n,k)$, for $1 \leq k \leq n$. These elements represent the column coefficients, which are propagated horizontally (direction of $(0,1,0)$). The computation of the new matrix is specified by:

$$\text{(variable } V\text{)}$$
$$\mathbf{E}_{10} = (D_{5,1}, V, in_{V3})$$
$$\mathbf{E}_{11} = (D_{5,2}, V, (V, H, RC), update, (\mathcal{I}_1, \mathcal{I}_2, \mathcal{I}_3))$$
$$\mathbf{E}_{12} = (D_6, V, V, id, \mathcal{I}_1)$$
$$\text{(variable } H - \text{column coefficients)}$$
$$\mathbf{E}_{13} = (D_6, H, V, id, \mathcal{I}_1)$$
$$\mathbf{E}_{14} = (D_7, H, H, id, \mathcal{I}_2)$$
$$\text{(variable } RC - \text{row coefficients)}$$
$$\mathbf{E}_{15} = (D_{5,1}, RC, (V, H), div, (\mathcal{I}_1, \mathcal{I}_2))$$
$$\mathbf{E}_{16} = (D_{5,2}, RC, RC, id, \mathcal{I}_3)$$

where:

- the index mappings are as above;

- the applied functions are:

$$in_{V3}(i, j, k) = 0$$
$$id(a) = a$$
$$div(a, b) = a/b$$
$$update(a, b, c) = a - b * c$$

- the domains are:

$$D_{5,1} = \{(i, j, k) \mid i = n, n + 1 \leq j \leq 2n - 1, k = 1\}$$
$$D_{5,2} = \{(i, j, k) \mid i = n, n + 1 \leq j \leq 2n - 1, 2 \leq k \leq n\}$$
$$D_6 = \{(i, j, k) \mid i = n, j = n, 1 \leq k \leq n\}$$
$$D_7 = \{(i, j, k) \mid i = n, n + 1 \leq j \leq 2n - 1, 1 \leq k \leq n\}.$$

The data dependence graph is given in Fig. 5.30 a) and b), and for $k = 1$ in Fig. 5.31. The iteration itself is illustrated in Fig. 5.32.

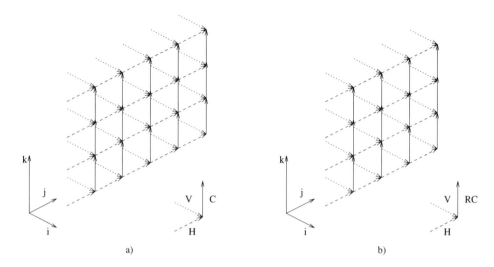

Fig. 5.30. Data dependence graph: a) finding the pivot $(i = c$ and $1 \leq c \leq n - 1)$; b) computing the new matrix $(i = n)$.

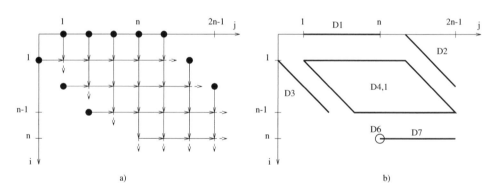

Fig. 5.31. Iteration for $k = 1$: a) data dependence graph; b) domains.

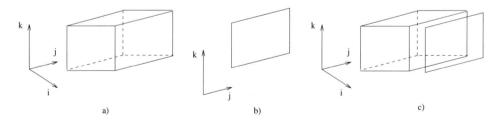

Fig. 5.32. Phases of the iteration: a) finding the pivot; b) computing the new matrix; c) composition of the phases.

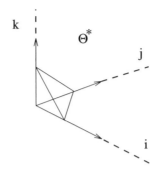

Fig. 5.33. Dependence Cone.

5.4.3   Space-Time Mapping

At each iteration t for $1 \leq t \leq n-1$, a submatrix of $(n+1-t) \times (n+1-t)$ elements is considered, as the remaining $t-1$ rows and columns of the current matrix are already in their final format. On each of these submatrices the same sequence of operations is performed. Therefore the specification given in the previous section can be used as the specification of each iteration of the algorithm (indeed the computation domains have to be "resized" accordingly at each iteration). In order to obtain an array design for the algorithm, we first apply a space-time mapping to each iteration and compose the mapped iterations subsequently. Indeed, the combinations of different mappings and compositions of the iterations produce several distinct array designs for the algorithm.

The data dependencies of the specification are uniform and the corresponding dependence cone is

$$\Theta^* = cone(\{(1,0,0),(0,1,0),(0,0,1)\}).$$

The dependence cone, illustrated in Fig. 5.33, is pointed.

A valid affine timing function is determined by any $\lambda = (\lambda_1, \lambda_2, \lambda_3) \in \mathbf{Z}^3$ such that the following system of inequalities is satisfied:

$$\begin{aligned} \lambda_1 &> 0 \\ \lambda_2 &> 0 \\ \lambda_3 &> 0. \end{aligned}$$

The vector $\lambda = (1,1,1)$ is a possible choice, determining the timing function $t(i,j,k) = i+j+k$. With this timing function, a single iteration on an $n \times n$ matrix requires $O(4n)$ steps.

d	$\lambda \cdot d$	$a(d)$
$(1,0,0)$	1	$(1,0)$
$(0,1,0)$	1	$(0,0)$
$(0,0,1)$	1	$(0,1)$

Table 5.12. Transformation of data dependence vectors under the space-time mapping $[t,a]$.

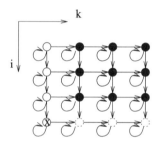

Fig. 5.34. Signal flow graph of the iteration.

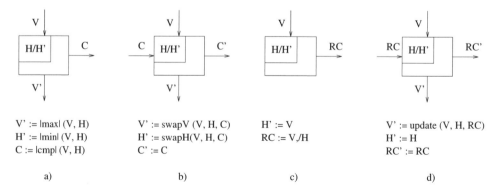

V' := |max| (V, H)
H' := |min| (V, H)
C := |cmp| (V, H)

a)

V' := swapV (V, H, C)
H' := swapH(V, H, C)
C' := C

b)

H' := V
RC := V,/H

c)

V' := update (V, H, RC)
H' := H
RC' := RC

d)

Fig. 5.35. Processing elements.

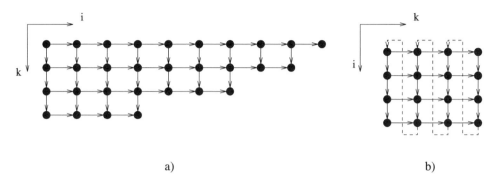

a)   b)

Fig. 5.36. Signal flow graphs: a) Sequential composition b) Superimposition.

A compatible allocation is determined by any projection vector $u = (u_1, u_2, u_3) \in \mathbf{Z}^3$ such that $\lambda \cdot u \neq 0$, that is $u_1 + u_2 + u_3 \neq 0$. For instance, $u = (0, 1, 0)$ satisfies the condition and determines the linear scheduling $a(i, j, k) = (i, k)$. The corresponding projection, summarised in Table 5.12, produces the signal flow graph of Fig. 5.34. There are four types of nodes in the graph: type a) (represented as white nodes) corresponds to the computations to determine the pivot; type b) (black nodes) corresponds to the computations to rearrange the entries of the current matrix; type c) (crossed white node at the bottom left-hand corner) corresponds to the computations to determine the row coefficients; and type d) (white dotted nodes) corresponds to the computations to update the entries of the current matrix. A description of the corresponding processing elements and their operations is given in Fig. 5.35. Note that signal flow graph and processing elements refer to a single iteration.

A sequential composition of the iterations produces the signal flow graph in Fig. 5.36 a) (where we have omitted the loops at each node and different types of nodes are not distinguished). As the iterations of the algorithm are strictly sequential, a more compact graph (which yields a more efficient use of the processing elements) can be obtained by superimposing the iterations. The resulting signal flow graph is given in Fig. 5.36 b). Regular array designs corresponding to both these graphs were proposed by Megson in [Meg90]. These designs compute the algorithm in, respectively, $O(2n^2 + n)$ and $O(n^2 + 3n)$ time steps. Note that, as data are pipelined between consecutive iterations, some of the computation of the next iteration can begin before those of the current iteration have all been completed.

Finally, note that both designs require the specification of control signals (which for brevity we have omitted) in order to tag the beginning of each new iteration.

5.5   Summary

We have presented a number of case studies for the illustration of the techniques developed in the previous chapters of this book, and shown that some interesting problems from the literature, such as cyclic reduction techniques for the solution of tridiagonal systems or the knapsack problem, can be treated systematically with our methods.

From the work presented in this chapter we can make a number of observations. While, in general, some expertise and problem-specific knowledge is required from the designer both for the initial specification of an algorithm and the application of the subsequent transformations, synthesis methods provide guidelines to support the designer in his/her task as well as the mechanisation of some of the design steps. These guidelines include, for instance, conditions related to the scalability of the array design (i.e., the uniformity of the regularisation directions with respect to the size parameters of the problem) and the existence of an affine scheduling (i.e., guaranteeing that the dependence cone is pointed). For example, in the case of cyclic reduction for the solution of tridiagonal systems, we could identify the lack of scalability of our specification in the early phases of the design, by analysing the generators of the dependence cone. On the other hand, little support is provided by synthesis methods to indicate how a "better" specification can be chosen. For instance, in the same example, the transformation $\mathcal{T}$ we used to obtain the new specification was chosen arbitrarily, relying upon our knowledge of the regularisation techniques and computational model.

Characterising what makes a specification particularly suitable for regular array synthesis is a non trivial problem, and very little work exists in the literature on the subject (the problem is mentioned in [LeXu91]). It is, however, an important issue as sometimes reformulating a problem is necessary (for instance, because the synthesis method breaks down). This was the case in our example of an M-to-1 decimator, where neither integral nor affine regularisation techniques could be applied successfully. Note that the necessary reformulation was not just a simple syntactic manipulation of the recurrences, but was based on a totally different view of the algorithm.

Changing the specification was also necessary in our example of Gaussian elimination with pivoting, where we replaced a dynamic algorithm with a static specification. We have already discussed (in Chapter 4) how a transformation from dynamic to static is necessary for the synthesis of dynamic problems as regular arrays. However, the type of transformation that we have applied in the example appears to be more complex than those we have defined through our regularisation techniques in Chapter 4. In particular, the transformation of the example was applied to the algorithm as a whole rather than its single dynamic data dependencies (as we do with our method). The relation between complexity and scope of application of synthesis transformations was discussed in Chapter 1, where we stressed how synthesis methods have developed by favouring the simpler, locally applicable transformations. The limitations of this approach appear to be critical with dynamic problems, where more global (hence complex) transformations seem to be required (even our techniques require the redefinition of sub-systems of equations in order to make a dynamic data dependence uniform).

The two dynamic examples, knapsack problem and Gaussian elimination with pivoting, have been useful to highlight the type of problems characterised by finitely generated dynamic recurrences. In the knapsack problem the dynamicity is resolved as soon as the weight distribution is known. The control variables are assigned once and maintain their values throughout the algorithm. On the other hand, Gaussian elimination requires to compute a new pivot at each iteration, hence (the same) dynamic data dependence relations have to be resolved several times throughout the algorithm.

We may conclude that finitely generated recurrences characterise a restricted class of dynamic problems, and that a more general treatment of dynamic problems requires the consideration of specifications at a more global (and abstract) level. We will return to these points in the next final chapter.

Chapter 6

Conclusions

This book contains a comprehensive treatment of regularisation techniques in the context of synthesis methods for regular array processors. In particular, Chapter 2 introduces a unifying notation and revisits some of the background theory for regularisation. This is of particular value as the theory appears fragmented in the literature with no standard notation adopted.

Throughout the book we have demonstrated some of the major benefits of synthesis methods, in particular, that:

- they provide a disciplined and rigorous approach to algorithm analysis and design, as the various stages of the design process are clearly identified and supported by the methods, and formal notation and transformations guarantee that the approach is well-founded;

- they represent an effective way of engineering parallel algorithms, as computationally powerful methods can be developed based on the underlying mathematical model of affine Euclidean geometry with embedded lattice spaces.

We have also exposed some of the limitations of the synthesis methods, primarily their restricted scope of application, and the fact that, even for problems within their scope, synthesis techniques have not reached the maturity of a formal method which provides *best practice without expertise* [Hal95]. In particular, a fairly high level of expertise is required from the algorithm designer throughout the design process to guide the transformations and provide efficient solutions under the given problem constraints.

The work of this book has aimed at overcoming some of the limitations of the existing techniques by widening their applicability to more general classes of problems. In doing so, we have been concerned primarily with the provision of practical solutions and developed our techniques within the traditional mathematical framework of synthesis methods. Our approach consists of the identification of classes of non-affine problems, and their systematic transformations into regular problems to which classic mapping techniques can be applied. Issues of scheduling and placement are also taken into account.

A major achievement of this work is the characterisation and systematic treatment of classes of integral and dynamic problems, that we have shown to be of practical interest by considering case studies from the literature to which our techniques apply.

The engineering solutions we have proposed are novel and have no counterparts in the literature. This makes their assessment difficult, as we need a better insight in their applicability and effectiveness. In the following section we make a number of observations and highlight issues which required further investigation.

6.1 Further work

Although our techniques have the merit to show that the systematic synthesis of non-affine and dynamic problems is feasible, we cannot make any claim of optimality in the treatment of such problems. An obvious source of inefficiency both for integral and dynamic techniques is the computational overhead due to the way data routing is defined. Although we have partially addressed the problem through the provision of parametric uniformisation techniques, further work on the reduction of routing overhead would be beneficial. Note that, while regularising integral data dependencies amounts to the definition of a suitable routing system, the treatment of dynamic data dependencies involves the definitions of two classes of computations: those computations necessary to establish the data dependence relation at run-time, and those computations necessary to route the data accordingly once such a relation has been established. Therefore, more efficient regularisation techniques for dynamic data dependencies should not just contain the routing overhead, but also maximise the parallel execution of these two classes of computations. In our techniques, this is achieved by separating the computations for determining the coefficients of a dynamic index mapping from those for computing the new values, hereby increasing their potential for parallelism. In the case of Gaussian elimination with pivoting, a similar effect is obtained, at each iteration, by rearranging the entries of the current matrix while determining the pivot. Indeed, efficient regularisation techniques could be developed by considering special subclasses of problems. This type of approach is taken in [Fr-et-al93, Sw-et-al94] for so-called *piece-wise linear data dependencies*, a type of integral dependencies in which the index mappings define piece-wise linear transformations.

A major difference between transformations for integral (in general static) and dynamic problems is the need, in the latter case, to consider algorithm specifications more globally rather than applying the transformations to single data dependencies in isolation. The reason for such a difference is that the computations which establish the dynamic data dependencies at run-time have to be taken into account as part of the new algorithm specification. With our techniques this is achieved by combining the separation of some computations of the algorithm and their translation in the computation space. A limitation of our approach, formally captured by the requirement of separability of the computations (see Section 4.3.6), is that the computation domain of the non-uniform data dependence remains unchanged throughout the transformations. In other words, the techniques provide a systematic routing of the data between pairs of computation points which are fixed by the specification. This characteristic of our techniques is also shared by classic regularisation techniques for affine problems [QuVa89] from which our approach has developed. However, more powerful regularisation techniques could be developed which modify the original computation domains, so that separability is not necessary. For instance, such techniques may allow the domains to be expanded to include the (possibly interleaved) computations of both the coefficients of the index mapping and the routing of the data. Such an approach would be particularly beneficial for problems such as Gaussian elimination with pivoting, in which the determination of the pivot and the corresponding updating of the matrix entries are interleaved throughout the algorithm. Indeed, such techniques are likely to be based on non-affine domain transformations. Also, it is necessary to address the introduction of new non-uniform data dependencies in the specification due to the modification of the original computation domain.

Our work on dynamic problems raises a number of questions. First of all, it could be argued whether our definition of dynamic data dependence provides an adequate abstraction for dynamic problems of practical interest, or whether other and more general forms of dynamic problems should be considered (a taxonomy of static and dynamic problems based on data dependencies and task generation can be found in [Me-et-al95]). We have shown that our approach is general enough to characterise interesting problems from the literature. However, we need a better insight into its applicability. Also we could question whether the mathematical framework in which our approach has been developed is adequate. Our treatment of dynamic problems, even in the restricted connotation which we have considered, shows that the existing framework lacks of some of the necessary basic notions, such as that of input. It would be of value to investigate other mathematical notations, which may be more appropriate for the expression and treatment of dynamic problems.

Finally, there are a number of issues which have not been addressed in this work. Among them, whether the computability of integral and dynamic problems can be established. It is possible that, because of the relationships among the various types of recurrences, some of the results known for other classes of recurrences (see the overview in Section 2.5) may be generalised to integral and dynamic problems. Also, although we have stressed the importance of providing synthesis techniques which can be mechanised, we have not considered the issue explicitly. Tool support for regular array synthesis exists (see the overview in Section 2.5) including algorithms for the manipulation of the recurrences, their domains and data dependencies. Those algorithms are based on known techniques from linear algebra, computational geometry and linear programming (such as determining standard basis of vector spaces, computing the convex hull of sets of points or solving linear optimisation problems – see Appendix D and work in [Ner63, PrSh85, Sch86]). The same type of algorithms could be applied for the manipulation of the specifications as defined by our regularisation techniques, which could then be easily integrated with existing libraries of transformations for regular array synthesis.

Appendix A

Notation

[set theory]

R	real numbers
Q	rational numbers
Z	integer numbers
N	natural numbers
$\mathbf{N}^+$	positive integer numbers
$\mathbf{R}^n$	n-dimensional Euclidean space
$\mathbf{Z}^n$	n-dimensional Euclidean lattice space
$\mathcal{U}$	generic set
$\mathcal{P}(\mathcal{U})$	powerset of $\mathcal{U}$
$|\mathcal{U}|$	cardinality of $\mathcal{U}$
$\mathcal{U}^n$	n-fold Cartesian product of $\mathcal{U}$, i.e., $\mathcal{U} \times \ldots \times \mathcal{U}$ n times
$pr_i : \mathcal{U}^n \to \mathcal{U}$	i^{th} projection mapping, for $i = 1, \ldots, n$
$[\mathcal{U} \to \mathcal{U}]$	set of all mappings from $\mathcal{U}$ to $\mathcal{U}$
$\mathbf{u}$	n-tuple in $\mathcal{U}^n$
$\mathcal{S}(\mathbf{u})$	*support set* of $\mathbf{u}$, defined as $\mathcal{S}(\mathbf{u}) = \cup_{i=1}^n \{pr_i(\mathbf{u})\}$
f	generic function
$f(\mathcal{U})$	image of $\mathcal{U}$ under f, i.e., $\{f(u) \mid u \in \mathcal{U}\}$
$range_t(\mathcal{N})$	range of t over $\mathcal{N}$, i.e., $\{t(n) \mid n \in \mathcal{N}\}$
f^m	composition of f with itself m times, i.e., $f \circ \ldots \circ f$, m times

[linear and affine algebra]

$\mathbf{I}_n$	$n \times n$ identity matrix
M	generic matrix
M_j	j^{th} row of M

M^{-1}	inverse of M
M^t	transpose of M
$null(M)$	null space of M
e_j	j^{th} vector in the standard basis of $\mathbf{R}^n$
$[\pi : \theta]$	hyperplane $\mathbf{R}^n$ with normal vector π and coefficient θ
D	generic set in $\mathbf{R}^n$
$lin(D)$	direction of D
$aff(D)$	affine hull of D
$conv(D)$	convex hull of D
P	generic convex polyhedral set in $\mathbf{R}^n$
$vert(P)$	set of points generating P
$ray(P)$	set of directions generating P
C	generic convex polyhedral cone in $\mathbf{R}^n$
$d_1, \ldots, d_m$	m vectors in $\mathbf{R}^n$
$cone(\{d_1, \ldots, d_m\})$	the smallest convex polyhedral cone generated by $d_1, \ldots, d_m$
$\langle d_1, \ldots, d_m \rangle$	linear space generated by $d_1, \ldots, d_m$
S	generic space (or set) in $\mathbf{R}^n$
$S^{\perp}$	orthogonal space of S
$\mathcal{T}$	generic affine transformation in $\mathbf{R}^n$
$\mathcal{L}_{\mathcal{T}}$	linear part of $\mathcal{T}$

[basic concepts in regular array synthesis]

$\mathcal{CS}$	computation space
$\mathcal{PS}$	processors space
Var	universe of variables
Val	universe of data values
$\mathcal{I}$	generic index mapping
$\Theta_{\mathcal{I}}$	dependence mapping defined by $\mathcal{I}$
$\Omega_{\mathcal{I}}$	dependence domain defined by $\mathcal{I}$
$\Theta_{\mathcal{I}}^*$	dependence cone defined by $\mathcal{I}$
$\perp$	undefined value (for variables)

[equations and systems]

$\mathbf{E}$	generic recurrence or input equation
$D_{\mathbf{E}}$	computation domain of $\mathbf{E}$
$^{\bullet}\mathbf{E}$	result of $\mathbf{E}$
$\mathbf{E}^{\bullet}$	arguments of $\mathbf{E}$
$f_{\mathbf{E}}$	applied function of $\mathbf{E}$

$\mathcal{IM}_{\mathbf{E}}$	index mappings of $\mathbf{E}$
$Var_{\mathbf{E}}$	set of variables of $\mathbf{E}$
$\mathbf{S}$	generic system of equations
$D_{\mathbf{S}}$	domain of $\mathbf{S}$
$Var_{\mathbf{S}}$	set of variables of $\mathbf{S}$
V	generic variable
$Def\mathbf{E}_V$	definition equations of V
$DefD_V$	definition domain of V
$Def\mathbf{S}_V$	definition subsystem of V

[data dependencies and graphs]

$\mathcal{DD}$	generic data dependence
$\mathcal{DD}_{\mathbf{E}_i}$	i^{th} data dependence of $\mathbf{E}$
$\mathcal{DD}_{\mathbf{E}}$	data dependencies of $\mathbf{E}$
$\mathcal{DD}_{\mathbf{S}}$	data dependencies of $\mathbf{S}$
$\mathcal{DDG}$	generic data dependence graph
$\mathcal{CDDG}$	generic complete data dependence graph
$\mathcal{RDG}$	generic reduced dependence graph
$\mathcal{EDG}$	generic extended dependence graph

[timing and allocation functions]

t	generic timing function
λ	vector defining of an affine timing function
μ	coefficient of an affine timing function
a	generic allocation function
σ	matrix defining of an affine allocation function
$[t, a]$	space-time mapping determined by t and a
$G^{t,a}$	signal flow graph under t and a

[miscellaneous]

tr	translation in $\mathbf{R}^n$
ren	variable renaming
$\mathbf{S}^{tr,ren}$	translated image of $\mathbf{S}$
in	generic input
$R, R^1, \ldots, S, S^1, \ldots$	routing variables
$\alpha, \beta, \gamma, \ldots$	control variables
$G_1, \ldots, G_m$	integer-valued variables
$g_1, \ldots, g_m$	integer functions
$d, \hat{d}, \check{d}$	regularisation direction vectors
$r, \hat{r}, \check{r}$	rays of a dependence cone
mod	modulus function

$\lfloor \_ \rfloor$ floor function ($\_$ place holder)

∞ infinity

p parameter

For simplicity, we usually represent an n-dimensional vector as an n-tuple, that is $x = (x_1, \ldots, x_n)$. The only exception to this convention is the case of matrix expressions, in which vectors are represented as column vectors.

As we restrict ourselves to lattice spaces, with some abuse of notation, we represent vector spaces, such as $lin(D)$ or $null(M)$, as lattice spaces.

Appendix B

Graph Theory

This appendix is based on [Car79].

B.1   Graphs

A graph is a pair $\mathcal{G} = (\mathcal{N}, \mathcal{A})$ such that $\mathcal{N}$ is a finite set and $\mathcal{A} \subseteq \mathcal{N} \times \mathcal{N}$. The set $\mathcal{N}$ is called the set of nodes of $\mathcal{G}$ and $\mathcal{A}$ its a set of arcs. The empty graph is the graph $\mathcal{G} = (\emptyset, \emptyset)$.

A graph $\mathcal{G}$ is said to be simple if and only if

- for all $n_i \in \mathcal{N}$, $(n_i, n_i) \notin \mathcal{A}$; and

- $(n_i, n_j) \in \mathcal{A}$ implies $(n_j, n_i) \in \mathcal{A}$.

Let $\mathcal{G} = (\mathcal{N}, \mathcal{A})$ be a graph and $\mathcal{P} = \{\mathcal{N}_1, \ldots, \mathcal{N}_n\}$ be a partition of $\mathcal{N}$. The condensation of $\mathcal{G}$ induced by P is the graph $\mathcal{G}_\mathcal{P} = (\mathcal{P}, \mathcal{A}_\mathcal{P})$ such that $\mathcal{A}_\mathcal{P} = \{(\mathcal{N}_r, \mathcal{N}_s) \in \mathcal{P} \times \mathcal{P} \mid \mathcal{N}_r \neq \mathcal{N}_s$ and $\exists n_i \in \mathcal{N}_r, n_j \in \mathcal{N}_s$ such that $(n_i, n_j) \in \mathcal{A}\}$.

The simplification of a graph $\mathcal{G}$ is the graph $\mathcal{G}_\mathcal{S} = (\mathcal{N}, \mathcal{A}_\mathcal{S})$ such that $\mathcal{A}_\mathcal{S} = \{(n_i, n_j) \in \mathcal{N} \times \mathcal{N} \mid n_i \neq n_j$ and either $(n_i, n_j) \in \mathcal{A}$ or $(n_j, n_i) \in \mathcal{A}\}$.

A path p of a graph $\mathcal{G}$ is a finite sequence $(n_0, n_1), (n_1, n_2), \ldots, (n_{r-1}, n_r)$ such that $\forall i, j, (n_i, n_j) \in \mathcal{A}$. The order of p is the number of arcs in the sequence. A path p is called a cycle if $n_0 = n_r$. A loop is a cycle of order one.

Let $\mathcal{G} = (\mathcal{N}, \mathcal{A})$ be a graph and $n \in \mathcal{N}$. Then:

- a descendant of n is a node n' such that there exists a path (n_0, n_1), $(n_1, n_2), \ldots, (n_{r-1}, n_r)$ with $n_0 = n$ and $n_r = n'$. We denote by $n_\downarrow$ the set of all the descendants of n; and

– a node accessible from n is a node n' such that $n' \in n_{\downarrow}$ or $n' = n$. We denote by $n^{\mathcal{A}}$ the set of all nodes accessible from n.

B.2 Connectivity Relations

Let $\mathcal{G} = (\mathcal{N}, \mathcal{A})$ be a graph. Then:

– the connectivity relation $\triangleright \subseteq \mathcal{N} \times \mathcal{N}$ of $\mathcal{G}$ is the equivalence relation such that $n \triangleright m$ if and only if $m \in n^{\mathcal{A}}$ on the simplification $\mathcal{G}_{\mathcal{S}}$ of $\mathcal{G}$. When $n \triangleright m$, n is said to be connected to m; and

– the connected components of $\mathcal{G}$ are the subgraphs of $\mathcal{G}$ generated by the equivalence classes of $\triangleright$. If $\mathcal{G}$ has only one connected component, $\mathcal{G}$ is said to be connected.

– the strong connectivity relation $\bowtie \subseteq \mathcal{N} \times \mathcal{N}$ of $\mathcal{G}$ is the equivalence relation such that $n \bowtie m$ if and only if $m \in n^{\mathcal{A}}$ and $n \in m^{\mathcal{A}}$ on $\mathcal{G}$. When $n \bowtie m$, n is said to be strongly connected to m; and

– the strongly connected components of $\mathcal{G}$ are the subgraphs of $\mathcal{G}$ generated by the equivalence classes of $\bowtie$. If $\mathcal{G}$ has only one strongly connected component, $\mathcal{G}$ is said to be strongly connected.

Let $\mathcal{G} = (\mathcal{N}, \mathcal{A})$ be a graph and $\bowtie$ its strong connectivity relation. The reduced graph $\mathcal{G}_{\mathcal{R}}$ of $\mathcal{G}$ is the condensation of $\mathcal{G}$ induced by $\mathcal{N} / \bowtie$.

B.3 Graph Operations

Let $\mathcal{G} = (\mathcal{N}, \mathcal{A})$ be a graph and $\mathcal{N}' \subseteq \mathcal{N}$. The restriction of $\mathcal{G}$ to $\mathcal{N}'$ is the graph $\mathcal{G}|_{\mathcal{N}'} = (\mathcal{N}', \mathcal{A}')$ such that $\mathcal{A}' = \{(n, n') \in \mathcal{A} \mid n, n' \in \mathcal{N}'\}$.

Let $\mathcal{G}_j = (\mathcal{N}_j, \mathcal{A}_j)$ be graphs, for $j = 1, \ldots, r$. Their union is the graph $\mathcal{G} = (\mathcal{N}, \mathcal{A})$ such that $\mathcal{N} = \cup_j \mathcal{N}_j$ and $\mathcal{A} = \cup_j \mathcal{A}_j$.

Appendix C

Convex Sets and Polyhedra

This appendix is based on [Roc70, Sch86]. In the appendix, for $x = (x_1, \ldots, x_n), y = (y_1, \ldots, y_n) \in \mathbf{R}^n$, $x \cdot y$ denotes their scalar product $x_1 y_1 + \ldots + x_n y_n$.

C.1 Combinations

Let $x_1, \ldots, x_m$ be m points in $\mathbf{R}^n$. A vector sum $\lambda_1 x_1 + \ldots + \lambda_m x_m$, with $\lambda_1, \ldots, \lambda_m \in \mathbf{R}$, is called:

- a linear combination of $x_1, \ldots, x_m$;

- an affine combination of $x_1, \ldots, x_m$, if $\lambda_1 + \ldots + \lambda_m = 1$;

- a convex combination of $x_1, \ldots, x_m$, if the coefficients λ_i are all non-negative and $\lambda_1 + \ldots + \lambda_m = 1$;

- a positive (non-negative) linear combination of $x_1, \ldots, x_m$, if the coefficients λ_i are all positive (non-negative);

- an integer combination of $x_1, \ldots, x_m$, if the coefficients λ_i are all integers.

A vector sum $\lambda_1 x_1 + \ldots + \lambda_k x_k + \lambda_{k+1} x_{k+1} \ldots + \lambda_m x_m$, is called a convex combination of m points and directions if all the coefficients λ_i are non-negative and $\lambda_1 + \ldots + \lambda_k = 1$ for a fixed k, with $0 \leq k \leq m$ ($k = 0$ means that there is no requirement about certain coefficients adding up to 1).

C.2   Affine Sets and Transformations

If x and y are distinct points in $\mathbf{R}^n$, the set of points of the form $(1-\lambda)x+\lambda y$, for $\lambda \in \mathbf{R}$, is called the line through x and y.

A subset M of $\mathbf{R}^n$ is called an affine set if it contains the line through any pair of its points, i.e., if $(1-\lambda)x + \lambda y \in M$ for every $x \in M$, $y \in M$ and $\lambda \in \mathbf{R}$. The empty set $\emptyset$, $\mathbf{R}^n$ and all singleton sets are affine.

For $M \subset \mathbf{R}^n$ and $a \in \mathbf{R}^n$, the translate of M by a is $M + a = \{a + x \mid x \in M\}$. An affine set M is said to be parallel to an affine set L if $M = L+a$ for some a.

The subspaces of $\mathbf{R}^n$ are the affine sets which contain the origin.

Each non-empty affine set M is parallel to a unique subspace L, given by $L = M - M = \{x - y \mid x \in M, y \in M\}$. L is called the direction of M and is denoted by $lin(M)$.

The dimension $dim(M)$ of an affine set M is defined as the dimension of the subspace L parallel to it, i.e., $dim(M) = dim(lin(M))$. By convention, $dim(\emptyset) = -1$. Affine sets of dimension $0, 1$ and 2 are called points, lines and planes, respectively. An $(n-1)$-dimensional affine set in $\mathbf{R}^n$ is called a hyperplane.

Given $\beta \in \mathbf{R}$ and a non-zero $b \in \mathbf{R}^n$, the set $H = \{x \mid b^t \cdot x = \beta\}$ is a hyperplane in $\mathbf{R}^n$. Every hyperplane may be represented in this way, with b and β unique up to a common non-zero multiple. b is called a normal to the hyperplane H. We denote a hyperplane H by $[b : \beta]$.

Given $b \in \mathbf{R}^m$ and an $m \times n$ real matrix B, the set $H = \{x \in \mathbf{R}^n \mid Bx = b\}$ is an affine set in $\mathbf{R}^n$. Every affine set may be represented in this way. (Therefore, any affine subset of $\mathbf{R}^n$ is an intersection of a finite collection of hyperplanes.)

The intersection of an arbitrary collection of affine sets is affine. Given $S \in \mathbf{R}^n$, the intersection of all the affine sets containing S is called the affine hull of S and is denoted by $aff(S)$. $aff(S)$ is the unique smallest affine set containing S and consists of all the affine combinations of the elements of S, i.e., $aff(S) = \{\sum_i \lambda_i x_i \mid x_i \in S, \sum_i \lambda_i = 1\}$.

A set of $m+1$ points $b_0, b_1, \ldots, b_m$ is said to be affinely independent if $aff(\{b_0, b_1, \ldots, b_m\})$ is m-dimensional. Therefore $b_0, b_1, \ldots, b_m$ are affinely independent if and only if $b_1 - b_0, \ldots, b_m - b_0$ are linearly independent.

The affine transformations from $\mathbf{R}^n$ to $\mathbf{R}^m$ are the mappings $\mathcal{T}$ of the form $\mathcal{T}(x) = \mathcal{L}(x) + b$, where $\mathcal{L}$ is a linear transformation from $\mathbf{R}^n$ to $\mathbf{R}^m$ and $b \in \mathbf{R}^m$.

C.3   Convex Sets

If x and y are distinct points in $\mathbf{R}^n$, the set of points of the form $(1-\lambda)x+\lambda y$, for $0 \leq \lambda \leq 1$, is called the closed line segment between x and y.

A subset C of $\mathbf{R}^n$ is said to be convex if it contains the closed line segment between any two of its points, i.e., $(1-\lambda)x+\lambda y \in C$ for every $x \in M$, $y \in M$ and $0 < \lambda < 1$. The empty set $\emptyset$ and $\mathbf{R}^n$ are convex.

Given $\beta \in \mathbf{R}$ and a non-zero $b \in \mathbf{R}^n$, the sets $\{x \mid b^t \cdot x \leq \beta\}$ and $\{x \mid b^t \cdot x \geq \beta\}$ are called closed half-spaces, and the sets $\{x \mid b^t \cdot x < \beta\}$ and $\{x \mid b^t \cdot x > \beta\}$ are called open half-spaces. Half-spaces are non-empty and convex.

A subset of $\mathbf{R}^n$ is convex if and only if it contains all the convex combinations of its elements.

The intersection of an arbitrary collection of convex sets is convex. Given $S \in \mathbf{R}^n$, the intersection of all the convex sets containing S is called the convex hull of S and is denoted by $conv(S)$. $conv(S)$ is the unique smallest convex set containing S and consists of all the convex combinations of the elements of S, i.e., $conv(S) = \{\sum_i \lambda_i x_i \mid x_i \in S, \lambda_i \geq 0, \sum_i \lambda_i = 1\}$.

The dimension $dim(C)$ of a convex set C is defined as the dimension of the affine hull $aff(C)$ of C.

C.4   Cones

A subset K of $\mathbf{R}^n$ is called a cone if it is closed under non-negative scalar multiplication, i.e., $\lambda x \in K$ for every $x \in K$ and $\lambda \geq 0$. A convex cone is a cone which is convex.

Let K be a convex cone. Then there is a smallest subspace containing K, namely $K - K = \{x - y \mid x \in K, y \in K\} = aff(K)$, and there is a largest subspace contained within K, namely $(-K) \cap K$. Cones are not necessarily pointed. For instance, subspaces are in particular convex cones. A cone K is pointed if and only if $(-K) \cap K = \{0\}$.

C.5   Recession Cone and Unboundedness

Unbounded closed convex sets have a simple behaviour at infinity. If C is an unbounded closed convex set and $x \in C$, then C contains some entire half-lines starting at x. The directions of such half-lines do not depend on x:

the half-lines of C starting at a different point y are just translates of those starting at x.

The direction of the half-line $\{x + \lambda y \mid \lambda \geq 0\}$, where $y \neq 0$, is defined as the set of all translates of the half-line, and is independent of x. This is called the direction of y. Two vectors have the same direction if and only if they are positive scalar multiples of each other. The zero vector has no direction.

Let C be a non-empty convex set in $\mathbf{R}^n$. C recedes in the direction of y if C includes all the half-lines in the direction of y which start at points in C, i.e., C recedes in the direction of y, where $y \neq 0$, if and only if $x + \lambda y \in C$ for every $\lambda \geq 0$ and $x \in C$.

The recession cone of C, denoted by 0^+C, is the set of all vectors $y \in \mathbf{R}^n$ satisfying the latter condition, including $y = 0$. The directions of the recession of C are directions in which C recedes.

Let C be a non-empty convex set. The recession cone 0^+C is a convex cone. It is the same as the set of vectors y such that $C + y \subset C$.

A non-empty closed convex set C in $\mathbf{R}^n$ is bounded if and only if its recession cone 0^+C consists of the zero vector alone.

If C is a non-empty convex set, the set $(-0^+C) \cap 0^+C$ is called the linearity space of C. It consists of the zero vector and all the non-zero vectors y such that, for every $x \in C$, the line through x in the direction of y is contained in C. The directions of the vectors y in the linearity space are called directions in which C is linear. If the linearity space has dimension 0, C is called pointed.

C.6   Polyhedral Convex Sets

A polyhedral convex set in $\mathbf{R}^n$ is a set which can be expressed as the intersection of some finite collection of closed half-spaces. Every affine set is polyhedral.

A polyhedral convex cone in $\mathbf{R}^n$ is a set which can be expressed as the intersection of a finite collection of closed half-spaces whose boundary hyperplanes pass through the origin.

A finitely generated convex set is a set which is the convex hull of a finite set of points and directions. Thus C is a finitely generated convex set if and only if there exist vectors $x_1, \ldots, x_m$ such that, for a fixed integer k, with $0 \leq k \leq m$, C consists of all the vectors of the form

$$x = \lambda_1 x_1 + \ldots + \lambda_k x_k + \lambda_{k+1} x_{k+1} \ldots + \lambda_m x_m$$

where λ_i are non-negative and $\lambda_1 + \ldots + \lambda_k = 1$.

The finitely generated convex sets which are cones are the sets which can be expressed this way with $k = 0$, i.e., with no requirement about certain coefficients adding up to 1. In such an expression, $\{x_1, \ldots, x_m\}$ is called a set of generators of the cone. A finitely generated convex cone is the convex hull of the origin and finitely many directions.

The cone generated by the vectors $x_1, \ldots, x_m$ is $cone(x_1, \ldots, x_m) = \{\lambda_1 x_1 + \ldots + \lambda_m x_m \mid \lambda_1, \ldots \lambda_m \geq 0\}$, i.e., it is the smallest convex cone containing $x_1, \ldots, x_m$.

The finitely generated convex sets which are bounded are called polytopes.

The property of being polyhedral is a finiteness condition on the external representations of a convex set. The property of being finitely generated is a finiteness condition on the internal representations of a convex set. The two properties are equivalent, and polyhedral convex sets are the same as finitely generated convex sets.

The concepts of polyhedron and polytope are related under the so-called decomposition theorem. The theorem states that a set P of vectors in a Euclidean space is a convex polyhedron if and only if $P = Q + C$ for some convex polytope Q and some polyhedral convex cone C.

Also, if $P = Q + C$ for some convex polytope Q and some polyhedral convex cone C, then C is the recession cone of P.

We say that P is generated by the points $x_1, \ldots, x_m$ and the directions $y_1, \ldots, y_t$ if

$$P = conv(\{x_1, \ldots, x_m\}) + cone(\{y_1, \ldots, y_t\}).$$

In this case we adopt the notation $vert(P)$ to denote the set $\{x_1, \ldots, x_m\}$ and $ray(P)$ for the set $\{y_1, \ldots, y_t\}$.

C.7 Duality

Let P be a set in $\mathbf{R}^n$. The dual (or polar) cone of P is the cone $\hat{P} = \{z \in \mathbf{R}^n \mid z \cdot x \geq 0, \forall x \in P\}$.

If P and P' are such that $P \subseteq P'$, then $\hat{P} \subseteq \hat{P}'$.

If C is a polyhedral convex cone, then $\hat{C}$ is also a polyhedral convex cone. If C is finitely generated by the set $\{r_i, \ldots, r_m\}$, then $\hat{C} = \{z \in \mathbf{R}^n \mid z \cdot r_1 \geq 0, \ldots, z \cdot r_m \geq 0\}$.

There exists a one-to-one correspondence between the generators of C and those of $\hat{C}$. In particular, let C be an n dimensional cone with n generators $\{r_i, \ldots, r_n\}$, and let Q denote a matrix having those generators as columns. Then the generators of $\hat{C}$ are the column of the matrix $\hat{Q} = -(Q^{-1})^t$, i.e., the opposite of the transpose of the inverse of Q.

C.8 Separating Hyperplane Theorem

If C is a finitely generated cone in $\mathbf{R}^n$ and d is a vector not in C, then there exists $\pi \in \mathbf{R}^n$ such that $\pi \cdot x \geq 0$ for all $x \in C$, and $\pi \cdot d < 0$. That is, the hyperplane $[\pi : 0]$ divides $\mathbf{R}^n$ into two parts, one containing d and the other containing C. In addition, C is pointed if and only if $\pi \cdot x > 0$ for all $x \in C$.

C.9 Other Results

Let $\mathcal{L}$ be a linear transformation from $\mathbf{R}^n$ to $\mathbf{R}^m$. Then $\mathcal{L}(C)$ is a polyhedral convex set in $\mathbf{R}^m$ for each polyhedral convex set C in $\mathbf{R}^n$, and $\mathcal{L}^{-1}(D)$ is a polyhedral convex set in $\mathbf{R}^n$ for each polyhedral convex set D in $\mathbf{R}^m$.

If D is a convex polyhedron and $y_1, \ldots, y_m$ are directions in $\mathbf{R}^n$, the polyhedron generated by D and $y_1, \ldots, y_m$ is defined as the set

$$P = \{z + \lambda_1 d_1 + \ldots + \lambda_m d_m \mid z \in D, \lambda_i \geq 0\}.$$

Appendix D

Aspects of Linear Algebra

This appendix is based on [Ner63, Ban93]. In this appendix, V denoted a generic vector space over a generic field F. If $v_1, \ldots, v_p$ are vectors in V, then $\langle v_1, \ldots, v_p \rangle$ denotes the subspace spanned by $v_1, \ldots, v_p$.

D.1 Elementary Row Operations and Elementary Matrices

There are three types of elementary row operations[1] defined on (the rows of) a matrix M: i) multiply a row of M by a non-zero scalar; ii) add a multiple of one row to another row; iii) interchange two rows.

An elementary matrix is any matrix obtained from an identity matrix by any elementary row operation. Therefore there is a one-to-one correspondence between elementary row operations and elementary matrices. An elementary matrix is a non-singular matrix and its inverse matrix is also an elementary matrix.

An elementary row operation on a matrix M can be accomplished by premultiplying M by the corresponding elementary matrix.

Any non-singular matrix M can be written as a product of elementary matrices.

---

[1]Elementary column operations are defined in a similar way. In this work we consider elementary row operations only.

D.2 Hermite Normal Form

A matrix in Hermite normal form[2] has the following form:

$$
\begin{bmatrix}
0 & \cdots & 0 & | & 1 & | & x & x & | & 0 & | & x & | & 0 & | & x & x \\
0 & \cdots & 0 & | & 0 & | & 0 & 0 & | & 1 & | & x & | & 0 & | & x & x \\
0 & \cdots & 0 & | & 0 & | & 0 & 0 & | & 0 & | & 0 & | & 1 & | & x & x \\
0 & \cdots & 0 & | & 0 & | & 0 & 0 & | & 0 & | & 0 & | & 0 & | & 0 & 0 \\
\vdots & \vdots & \vdots & | & \vdots & | & \vdots & \vdots & | & \vdots & | & \vdots & | & \vdots & | & \vdots & \vdots \\
0 & \cdots & 0 & | & 0 & | & 0 & 0 & | & 0 & | & 0 & | & 0 & | & 0 & 0
\end{bmatrix}
$$

where an x denotes any number.

Any $m \times n$ M matrix can be reduced to its Hermite normal form, and such a form is unique. The reduction to Hermite normal form can be achieved by a series of elementary row operations. As the form is unique, it is independent from the the particular sequence of operations chosen.

If M is a non-singular square matrix, its Hermite normal form is the identity matrix.

The Hermite normal form has a number of important applications in linear algebra among which finding a standard basis for a subspace S and its orthogonal complement $S^\perp$, determining the linearly independent vectors among the vectors of a set, solving systems of linear equations, inverting a matrix, etc. (see this appendix later on).

D.3 Integer Elementary Row Operations

As we mainly work in $\mathbf{Z}^n$, then in general, we restrict ourselves to integer matrices and integer elementary row operations [Ban93] There are three types of integer elementary row operations: i) multiply a row of M by -1; ii) add an integer multiple of one row to another row; iii) interchange two rows.

D.4 Unimodularity

A square integer matrix M is unimodular if its determinant $det(M)$ is equal to ±1.

Each unimodular matrix is the result of a finite sequence of integer elementary row operations performed on the identity matrix (of the same size).

---

[2]This form is often called row-echelon form.

Unimodular matrices have the following properties: the transpose and the inverse of a unimodular matrix is unimodular; the product of two unimodular matrices is unimodular.

D.5   Echelon Form

Let M be an $m \times n$ integer matrix and l_i denote the column number of the leading element of row i (for a zero row, l_i is undefined). Then M is in Echelon form[3] if for some integer ρ, with $0 \leq \rho \leq m$, the following conditions hold:

- rows from 1 to ρ are non-zero rows;

- rows from $\rho + 1$ to m are zero rows;

- for $1 \leq i \leq \rho$, each element in column l_i below row i is zero;

- $l_1 < l_2 < \ldots < l_\rho$.

The leading element of a row need not be equal to 1. A zero matrix is an echelon matrix for which $\rho = 0$. The rank of a matrix in Echelon form is equal to ρ, i.e., $rank(M) = \rho$.

A matrix in Echelon form has the following form:

$$\begin{bmatrix} 0 & \cdots & 0 & | & x_1 & | & x & x & | & x & | & x & | & x & | & x & x \\ 0 & \cdots & 0 & | & 0 & | & 0 & 0 & | & x_2 & | & x & | & x & | & x & x \\ 0 & \cdots & 0 & | & 0 & | & 0 & 0 & | & 0 & | & 0 & | & x_\rho & | & x & x \\ 0 & \cdots & 0 & | & 0 & | & 0 & 0 & | & 0 & | & 0 & | & 0 & | & 0 & 0 \\ \vdots & \vdots & \vdots & | & \vdots & | & \vdots & \vdots & | & \vdots & | & \vdots & | & \vdots & | & \vdots & \vdots \\ 0 & \cdots & 0 & | & 0 & | & 0 & 0 & | & 0 & | & 0 & | & 0 & | & 0 & 0 \end{bmatrix}$$

where x, x_i denote integer numbers.

Any integer matrix can be reduced to Echelon form by a sequence of integer elementary row operations. Applying a finite sequence of elementary row operations to a matrix is equivalent to premultiplying the matrix by a suitable unimodular matrix.

---

[3]In classical text of linear algebra, Hermite normal form and row-echelon form are exactly the same. The distinction we have made here is a little arbitrary. It would be more appropriate to call this form an integer Hermite normal form. However we adopt the terminology of [Ban93].

D.6 Linear Functional and Duality

Let V be a vector space over a field of constants F. A linear transformation $\mathcal{L}$ of V into F is called a linear form or linear functional on V. The set of all linear functionals on V is a vector space, called the dual or conjugate space of V, and denoted by $\hat{V}$. Besides these spaces have the same number of dimensions, i.e., $dim(\hat{V}) = dim(V)$.

If $B = \{b_1, \ldots, b_n\}$ is a basis of V then there exists a corresponding dual basis $\hat{B} = \{\hat{b}_1, \ldots, \hat{b}_n\}$ of $\hat{V}$. The relation between the two basis is characterised by the equations $\hat{b}_i(b_j) = \delta_{i,j}$, for all $i, j = 1, \ldots, n$, where $\delta_{i,j}$ is the Kronecker delta ($\delta_{i,j} = 1$ if $i = j$ and $\delta_{i,j} = 0$ if $i \neq j$).

Let $B' = \{b_1', \ldots, b_n'\}$ be another basis of V. Let $P = [p_{i,j}]$ be the matrix of transition from B to B', i.e., $b_i' = \sum_{i=1}^{n} p_{i,j} b_i$. Then P^t is the matrix of transition from basis $\hat{B}'$ to $\hat{B}$. Therefore $(P^t)^{-1} = (P^{-1})^t$ is the matrix of transition from $\hat{B}$ to $\hat{B}'$.

D.7 Annihilators

Let V be an n-dimensional vector space and $\hat{V}$ its dual. If for $v \in V$ and $\hat{v} \in \hat{V}$, we have $v \cdot \hat{v} = 0$, we say that v and $\hat{v}$ are orthogonal.

Let W be a subset of V. The set of all the linear functional $\hat{v}$ such that $\hat{v} \cdot w = 0$ for all $w \in W$, is called the annihilator or orthogonal complement of W, denoted by $W^{\perp}$. Any element of $W^{\perp}$ is called an annihilator of W. $W^{\perp}$ is a sub-space of $\hat{V}$. Besides if W is a sub-space of dimension ρ, then $\hat{V}$ is a sub-space of dimension $n - \rho$.

If W_1 and W_2 are two subspaces of V and $W_1^{\perp}$ and $W_2^{\perp}$ are their annihilators in $\hat{V}$, then the annihilator of $W1 + W_2$ is $W_1^{\perp} \cap W_2^{\perp}$, and the annihilator of $W1 \cap W_2$ is $W_1^{\perp} + W_2^{\perp}$.

D.8 Algorithmic Issues

D.8.1 Standard Basis of V

Let $B = \{b_1, \ldots, b_n\}$ be a basis of a vector space V. A standard basis of V can be found according to the following. Represent the basis as a matrix $B = [b_{i,j}]$ where each row is a vector of the basis. Reduce B to its Hermite normal form B'. Then the rows of B' represent a standard basis for V. This basis is standard because any basis of V reduces to such a basis by Hermite normal form reduction.

Set Spanning the Same Sub-space

To decide whether two sets of vectors span the same subspace, we can simply reduce the corresponding matrices to their Hermite normal forms and compare their rows.

Linear Independence

Given a set of vectors $\{x_1, \ldots, x_k\}$ in an n-dimensional space, their linear independence can be checked by defining a matrix M having such vectors as columns and reducing M to Hermite normal form. The standardised columns (those with just one non-null entry equal to 1) of the resulting matrix correspond to the linearly independent elements of the set.

D.8.2   Basis of $W_1 + W_2$

Given two subspaces W_1, W_2 of a linear space V, their sum is defined as $W_1 + W_2 = \{w_1 + w_2 \mid w_1 \in W_1, w_2 \in W_2\}$.

If B_1 and B_2 are bases of W_1 and W_2, respectively, then $B_1 \cup B_2$ spans $W_1 + W_2$. Finding a basis for $W_1 + W_2$ amounts to discarding the dependent vectors of $B_1 \cup B_2$ until an independent spanning set remains. Therefore we can construct a matrix whose rows are the vectors in $B_1 \cup B_2$ and reduce it to Hermite normal form. The resulting rows constitute a basis of $W_1 + W_2$.

D.8.3   Basis of $W_1 \cap W_2$

Instead of finding the intersection $W1 \cap W_2$ directly, it is easier to find $W_1^\perp$ and $W_2^\perp$, then $W_1^\perp + W_2^\perp$ and finally $(W_1^\perp + W_2^\perp)^\perp$.

D.8.4   Basis of $W^\perp$

Given a basis $\{b_1, \ldots, b_m\}$ of W, a basis of $W^\perp$ is a basis of the solution space of the homogeneous system of equations $B \cdot y = 0$, where B is the matrix having the vectors b_j as columns. This can be obtained by reducing B to Hermite normal form and finding the k independent columns of the resulting matrix, with $k \leq n$. If $k = n$ then W is full-dimensional and $W^\perp = \{0\}$. Otherwise, $W^\perp$ is a subspace of dimension $n - k$. A basis for such subspace can be found by arbitrarily considering $n - k$ variables as parameters and solving, with respect to them, the subsystem of equations resulting from the Hermite normal form of B.

Basis of *null(A)*

Finding a basis for $null(A)$, where A is an $n \times n$ matrix, is equivalent to providing a basis for the solution space of the homogeneous system of linear equations $A \cdot x = 0$ for $x \in \mathbf{R}^n$. Therefore, finding a basis of $null(A)$ is an instance of the problem of finding a basis for the orthogonal complement of a given subspace.

Bibliography

[Abd90]   P.Abdulla, *Decision problems in systolic circuits verification.* Uppsala University, PhD Thesis, 1990.

[Abd92]   P.Abdulla, "Automatic verification of a class of systolic circuits", *Formal Aspects of Computing*, vol. 4, pp. 149-194, 1992.

[Ada68]   D.A.Adams, "A computation model with data flow sequencing", *Stanford University, Technical Report*, no. CS-117, 1968.

[Ada70]   D.A.Adams, "A model for parallel computations", *Parallel Processor Systems, Technology and Applications*, L.C.Hobb *et al.* (eds.), pp. 311-333, 1970.

[AnIr91]   C.Ancourt, F.Irigoin, "Scanning polyhedra with DO loops", *Proceedings 3rd SIGPLAN Symposium on Principles and Practice of Parallel Programming*, pp. 39-50, ACM Press, 1991.

[An-et-al87] M.Annarote, E.Arnould, T.Gross, H.T.Kung, M.Lam, O.Menzilcioglu, J.A.Webb, "The Warp computer - architecture, implementation and performance", *IEEE Transactions on Computers*, vol. 36, no. 12, pp. 1523-1538, 1987.

[Bac78]   J.Backus, "Can programming be liberated from the Von Neumann style? A functional style and its algebra of programs", *Communication of ACM*, vol. 21, pp. 613-641, 1978.

[Ban93]   U.Banerjee, *Loop Transformations for Restructuring Compilers: The Foundations.* Kluwer Academic Publisher, 1993.

[Ban94]   U.Banerjee, *Loop Parallelization.* Kluwer Academic Publisher, 1994.

[BaEl88]   H.Barada, A.El-Amawy, "Systolic architecture for matrix trian-
              gularisation with partial pivoting", *IEE Proceedings*, vol. 135,
              no. 4, pp. 295-300, 1988.

[BaLe91]   M.Barnett, C.Lengauer, "A systolizing compilation scheme",
              University of Edinburgh, Tecnical Report, no. ECS-LFCS-91-
              134, January 1991.

[BaLe91b]   M.Barnett, C.Lengauer, "The synthesis of systolic programs",
              *Research Directions in High-Level Parallel Programming Lan-
              guages*, Lecture Notes in Computer Science, Springer-Verlag,
              no. 574, pp. 309-325, 1991.

[Bay94]   M.A.Bayoumi (ed.), *VLSI Design Methodologies for Digital
              Signal Processing Architectures*. Kluwer Academic Publishers,
              1994.

[Be-et-al90] A.Benaini, P.Quinton, Y.Robert, Y.Sauter, B.Tourancheau,
              "Synthesis of a new systolic architecture for the algebraic path
              problem", *Science of Computer Programming*, vol. 15, pp. 135-
              158, North-Holland, 1990.

[BoJe89]   G.S.Booles, R.C.Jeffrey, *Computability and Logic*. Open Uni-
              versity Set Book, Cambridge University Press, Third Edition,
              1989.

[Br-et-al87] W.Brauer, W.Reisig, G.Rozenberg (eds.), *Petri Nets: Central
              Models and their Properties*. Lecture Notes in Computer Sci-
              ence, nos. 254-255, Springer Verlag, 1980.

[Bro83]   A.Brøndsted, *An Introduction to Convex Polytopes*. Graduate
              Texts in Mathematics, Springer-Verlag, 1983.

[BuDe88]   J.C.Bu, E.F.Deprettere, "Converting sequential iterative algo-
              rithms to recurrent equations for automatic design of systolic
              arrays", *Proceedings IEEE International Conference on Acous-
              tics, Speech and Signal Processing (ICASSP 88), Vol. IV, VLSI:
              Spectral Estimation*, IEEE Press, pp. 2025-2028, 1988.

[Bu-et-al90] J.C.Bu, E.F.Deprettere, P.Dewilde, "A design methodology for
              fixed-size systolic arrays", *Proceedings International Conference
              on Application Specific Array*, IEEE Press, pp. 591-602, 1990.

[Bun83]   A.Bundy. *The computer modelling of mathematical reasoning.* Academic Press Inc., 1983.

[BuFa93]   R.L.Burden, J.D.Faires, *Numerical Analysis.* PWS Publishing, 1993.

[CaSt84]   P.Cappello, K.Steiglitz, "Unifying VLSI array design with linear transformations of space-time", *Advances in Computing Research*, no. 2, pp. 23-65, 1984.

[Car79]   B.Carré, *Graphs and Networks.* Oxford Applied Mathematics and Computing Science Series, Oxford University Press, 1979.

[CaGe89]   N.Carriero, D.Gelernter, "Linda in context", *Communications of ACM*, vol. 32, no. 4, pp. 444-458, 1989.

[ChKa70]   D.R.Chand, S.S.Kapur, "Algorithm for convex polytopes", *Journal of ACM*, vol. 17, no. 1, pp. 78-86, 1970.

[ChMi88]   K.Chandy, J.Misra, *Parallel Program Design.* Addison-Wesley, 1988.

[Che83]   M.C.Chen, *Space-time algorithms: semantics and methodology.* California Institute of Technology, PhD Thesis, 1983.

[Che86]   M.C.Chen, "A design methodology for synthesizing parallel algorithms and architectures", *Journal of Parallel and Distributed Computing*, vol. 3, no. 4, pp. 461-491, 1986.

[Che86b]   M.C.Chen, "A parallel language and its compilation to multiprocessor machines", *Proceedings 13th Annual Symposium on POPL*, pp. 131-139, 1986.

[ClMo93]   P.Clauss, C.Mongenet, "Synthesis aspects in the design of efficient processor arrays from affine recurrence equations", *Journal of Symbolic Computation*, vol. 15, pp. 547-569, 1993.

[CollMD]   E.J.Borowski, J.M.Borwein, Dictionary of Mathematics, Collins, 1989.

[ChMe93]   X.Chen, G.M.Megson, "A methodology of partitioning and mapping for fixed-shape and given-mesh regular arrays", *The University of Newcastle upon Tyne, Computing Science, Technical Report Series*, no. 423, 1993.

[Che65]   N.V.Chernikova, "Algorithm for finding a general formula for
            the non-negative solutions of a system of linear inequali-
            ties", *U.S.S.R. - Computational Mathematics and Mathematical
            Physics*, vol. 5, pp. 228-233, 1965.

[Chv83]   V.Chvátal, *Linear programming*. W.H.Freeman and Company,
            1983.

[Co-et-al86] M.Cosnard, P.Quinton, Y.Robert, M.Tchuente (eds.), *Parallel
            Algorithms and Architectures*, North-Holland, 1986.

[Cri97]   J.Crichlow, *An introduction to distributed and parallel comput-
            ing*. Prentice-Hall, 1997.

[CrRa83]   R.E.Crochiere, L.R.Rabiner, *Multirate digital signal processing*.
            Signal Processing Series, Prentice-Hall, 1983.

[Cu-et-al83] K.Culik II, J.Gruska, A.Salomaa, "Systolic automata for VLSI
            on balanced trees", *Acta Informatica*, vol. 18, pp. 335-344, 1983.

[Cu-et-al84] K.Culik II, J.Gruska, A.Salomaa, "Systolic trellis automata,
            Part II", *International Journal of Computer Mathematics*,
            vol. 16, pp. 3-22, 1984.

[Dan55]   G.B.Dantzig, "Upper bounds, secondary constraints, and block
            triangularity in linear programming", *Econometrica, Journal of
            the Econometric Society*, vol. 23, pp. 174-183, 1955.

[Dar91]   A.Darte, "Regular partitioning for synthesising fixed-size sys-
            tolic arrays", *Journal of VLSI Integration*, vol. 12, pp. 293-304,
            1991.

[Da-et-al91] A.Darte, L.Khachiyan, Y.Robert, "Linear scheduling is nearly
            optimal", *Parallel Processing Letters*, vol. 1, no. 2, pp. 73-81,
            1991.

[DaRo94]   A.Darte, Y.Robert, "Constructive methods for scheduling uni-
            form loop nests", *IEEE Transactions on Parallel and Dis-
            tributed Systems*, vol. 5, no. 8, pp. 814-822, 1994.

[DeIp86]   J.-M.Delosme, I.Ipsen, "Systolic arrays synthesis: computabil-
            ity and time cones", in [Co-et-al86], pp. 295-312, 1986.

[DeIp87]   J.-M.Delosme, I.Ipsen, "Efficient systolic arrays for the solution of Toeplitz systems: an illustration of a methodology for the construction of systolic architectures in VLSI", in [Mo-et-al87], pp. 37-46, 1987.

[Den80]   J.B.Dennis, "Data Flow Supercomputers", *IEEE Computers*, no. 18, pp. 42-56, 1980.

[DeWe77]   J.B.Dennis, K.S.Weng, "Applications of data flow computation to the weather problem", *High Speed Computer and Algorithm Organization*, D.J.Kuck, D.H.Lawrie, A.Sameh (eds.), pp. 143-157, Academic Press, 1977.

[DuMe84]   D.E.Dudgeon, R.M.Mersereau, *Multidimensional Digital Signal Processing*. Signal Processing Series, Prentice-Hall, 1984.

[EkTu87]   S.M.Eker, J.V.Tucker, "Specification, derivation and verification of concurrent line drawing algorithms and architectures", *The University of Leeds, Centre for Theoretical Computer Science, Report*, no. 10.87, 1987.

[ElBa90]   A.El-Amawy, H.Barada, "Efficient linear and bilinear arrays for matrix triangularisation with partial pivoting", *IEE Proceedings*, vol. 137, no. 4, pp. 295-300, 1990.

[Eva91]   D.J.Evans (ed.), *Systolic Algorithms*, Topics in Computer Mathematics:3, Gordon and Breach Science Publishers, 1991.

[FaNa88]   E.Fachini, M.Napoli, "C-tree systolic automata", *Theoretical Computer Science*, vol. 56, pp. 155-186, 1988.

[Fea92a]   P.Feautrier, "Some efficient solutions to the affine scheduling problem, part I, one dimensional time", *Journal of Parallel Programming*, vol. 21, no. 5, pp. 313-348, 1992.

[Fea92b]   P.Feautrier, "Some efficient solutions to the affine scheduling problem, part II, multidimensional time", *Journal of Parallel Programming*, vol. 21, no. 6, pp. 389-420, 1992.

[Fea94]   P.Feautrier, "Towards automatic distribution", *Parallel Processing Letters*, vol. 4, no. 3, pp. 233-244, 1994.

[Fo-et-al88] J.A.B.Fortes, K.S.Fu, B.W.Wah, "Systematic design approaches for algorithmically specified arrays", *Computer Architecture: Concepts and Systems*, J.M.Milutinovic (ed.), pp. 454-494, North Holland, 1988.

[FoMo84] J.A.B.Fortes, D.Moldovan, "Data broadcasting in linearly scheduled array processors", *Proceedings 11th Annual Symposium on Computer Architecture*, pp. 224-231, 1984.

[FoMo85] J.A.B.Fortes, D.I.Moldovan, "Parallelism detection and transformation techniques useful for VLSI algorithms", *Journal of Parallel and Distributed Computing*, vol. 2, pp. 277-301, 1985.

[FoWa87] J.A.B.Fortes, B.W.Wah, "Systolic arrays - from concepts to implementation", *IEEE Computer*, vol. 20, no. 7, pp. 12-17, 1987.

[Fr-et-al93] F.H.M.Franssen, F.Balasa, M.F.X.B.Van Swaaij, F.V.M.Catthoor, H.J.De Man, "Modelling multidimensional data and control flow", *IEEE Transactions on VLSI Systems*, vol. 1, no. 3, pp. 319-327, 1993.

[Ga-et-al87] P.Gachet, B.Joinnault, P.Quinton, "Synthesizing systolic arrays using DIASTOL", in [Mo-et-al87], pp. 25-36, 1987.

[Ga-et-al88] P.Gachet, P.Quinton, C.Mauras, Y.Saouter, "Alpha du Centaur: a prototype environment for the design of parallel regular algorithms", *IRISA, Publication Interne*, no. 439, 1988.

[GaPe92] E.Gautrin, L.Perraudeau, "MADMACS: a tool for the layout of regular arrays." *IRISA, Publication Interne*, no. 641, 1992.

[Gel85] D.Gelernter, "Generative communication in Linda", *ACM Transactions on Programming Languages and Systems*, vol. 7, no. 1, pp. 80-112, 1985.

[Ge-et-al90] D.Gelernter, A.Nicolau, D.Padua (eds.), *Languages and compilers for parallel computing.* Research Monograph in Parallel and Distributed Computing, Pitman, The MIT Press, 1990.

[GeKu81] W.M.Gentleman, H.T.Kung, "Matrix triangularization", *SPIE Real Time Signal Processing IV*, vol. 298, pp. 19-26 1981.

[GiRy88]   A.Gibbons, W.Rytter, *Efficient Parallel Algorithms*. Cambridge University Press, Cambridge, 1988.

[Grü67]   B.Grünbaum, *Convex Polytopes*. Interscience Publishers, 1967.

[Gru84]   J.Gruska, "Systolic automata: power, characterisation, non-homogeneity", *Proceedings Mathematical Foundations of Computer Science (MFCS '84)*, M.P.Chytil, V.Koubek (eds.), Lecture Notes in Computer Science, no. 176, pp. 32-49, Springer-Verlag, 1984.

[Gru90]   J.Gruska, "Synthesis, structure and power of systolic computations", *Theoretical Computer Science*, no. 71, pp. 47-77, 1990.

[GuLi82]   L.J.Guibas, F.M.Liang, "Systolic stacks, queues and counters", *1982 Conference on Advanced Research in VLSI*, M.I.T, 1982.

[Hal95]   J.G.Hall, "Combining formal methods: the two button press case study", *Lecture at the Colloquium on Practical Application of Formal Methods*, IEE Computing and Control Division, Professional Group C1 (Software engineering), May 1995, Digest No: 1995/109, 1995.

[Har92]   D.Harel, *Algorithmics: The Spirit of Computing*. Addison-Wesley Publishing Company, 1992.

[Hen86]   M.Hennessy, "Proving systolic systems correct", *ACM Transactions on Programming Languages and Systems*, vol. 8, no. 3, pp. 344-387, 1986.

[HoTu94]   K.M.Hobley, J.V.Tucker, "Clocks, Retiming and Transformations of Synchronous Concurrent Algorithms", *Transformational Approaches to Systolic Design*, G.M.Megson (ed.), Chapman & Hall, 1994.

[Ho-et-al89]   B.Hochet, P.Quinton, Y.Robert, "Systolic Gaussian elimination over GF(p) with partial pivoting", *IEEE Transactions on Computers*, vol. 38, no. 9, pp. 1321-1324, 1989.

[Hu82]   T.C.Hu, *Combinatorial Algorithms*. Addison-Wesley Publishing Company, 1982.

[HuLe87]   C.-H.Huang, C.Lengauer, "The derivation of systolic implemen-
            tation of programs", *Acta Informatica*, vol. 24, no. 6, pp. 595-
            632, 1987.

[Hi-et-al90] P.Hilfinger, J.Rabaey, D.Genin, C.Scheers, H.De Man, "DSP
            specifications using the SILAGE language", *Proceedings IEEE
            International Conference on Acoustics, Speech and Signal Pro-
            cessing*, pp. 1057-1060, 1990.

[HwBr85]   K.Hwang, F.A.Briggs, *Computer architecture and parallel pro-
            cessing*. Computer Science Series, McGraw-Hill International
            Editions, 1985.

[IrTr88]   F.Irigoin, R.Triolet, "Supernode partitioning", *Proceedings 15th
            POPL*, San Diego, California, pp. 319-328, 1988.

[Ka-et-al67] R.M.Karp, R.E.Miller, S.Winograd, "The organization of com-
            putations for uniform recurrence equations", *Journal of the
            ACM*, vol. 14, no. 3, pp. 563-590, 1967.

[KeMc90]   K.Kennedy, K.S.McKinley, "Loop distribution with arbitrary
            control flow", *IEEE/ACM, Proceedings Supercomputing '90*,
            New York, 1990.

[Kri89]   E.V.Krishnamurthy, *Parallel processing*. International Com-
            puter Science Series, Addison-Wesley, 1989.

[HTKun82]   H.T.Kung, "Why systolic architectures?", *IEEE Computer*,
            vol. 15, no. 1, pp. 37-46, 1982.

[KuLe80]   H.T.Kung, C.E.Leiserson, "Systolic arrays (for VLSI)", *In-
            troduction to VLSI Systems*, C.Mead and L.Conway (eds.),
            sect. 8.3, pp. 271-292, Addison-Wesley, 1980.

[KuLi84]   H.T.Kung, W.T.Lin, "An algebra for VLSI algorithm design",
            Technical Report, Carnegie Mellon University, no. CMU-CS-84-
            100, 1984.

[KuWe85]   H.T.Kung, J.A.Webb, "Global operations on the CMU Warp
            Machine", *Proceedings of 1985 AIAA Computer in Aerospace V
            Conference*, American Institute of Aeronautics and Astronau-
            tics, pp. 209-218, 1985.

[SYKun88] S.Y.Kung, *VLSI array processors*. Information and System Sciences Series, Prentice Hall, 1988.

[Ku-et-al81] S.Y.Kung, K.S.Arun, D.V.Bhaskar Rao, Y.H.Hu, "A matrix data flow language/architecture for parallel matrix operations based on computational wavefront concept", *Proceedings CMU Conference on VLSI Systems Computations*, pp. 235-244, Computer Science Press, 1981.

[Ku-et-al82] S.Y.Kung, K.S.Arun, R.J.Gal-Ezer, D.V.Bhaskar Rao, "Wavefront array processor: language, architecture, and applications", *IEEE Transactions on Computers*, vol. C-351, no. 11, pp. 1054-1065, 1982.

[Lam74] L.Lamport, "The parallel execution of DO loops", *Communications of the ACM*, vol. 17, no. 2, pp. 83-93, 1974.

[Le-et-al89] P.Lee, J.Wu, A.Yang, K.Yip, "SYSDES: a systolic array automation design system", *Proceedings 4th SIAM Conference on Parallel Processing for Scientific Computing*, 1989.

[Lei81] C.E.Leiserson, *Area-efficient VLSI computation*, PhD Thesis, Carnegie-Mellon University, 1981.

[LeSa83] C.E.Leiserson, F.Saxe, "Optimizing synchronous systems", *Journal of VLSI and Computer Systems*, vol. 1, no. 1, pp. 41-67, 1983.

[Len90] C.Lengauer, "Code generation for a systolic computer", *Software - Practice and Experience*, vol. 20, no. 3, pp. 261-282, 1990.

[Le-et-al91] C.Lengauer, M.Barnett, D.G.Hudson, "Towards systolizing compilation", *Distributed Computing*, vol. 5, pp. 7-24, 1991.

[LeXu91] C.Lengauer, J.Xue, "Recent developments in systolic design", *University of Edinburgh, Technical Report*, no. ECS-LFCS-91-176, 1991.

[LiWa85] G.-J.Li, B.W.Wah, "The design of optimal systolic arrays", *IEEE Transactions on Computers*, vol. C-34, no. 1, pp. 66-77, 1985.

[LiBa94]   N.Ling, M.A.Bayoumi, "From architecture to algorithm - a formal approach", in [Meg94], pp. 242-295, 1994.

[Lis89]   B.Lisper, "Single-assignment semantics for imperative programs", *Parallel Architectures and Languages Europe (PARLE '89), Vol. II, Parallel Languages*, Lecture Notes in Computer Science, no. 366, pp. 321-334, Springer-Verlag, 1989.

[LoZa94]   J.López, E.L.Zapata, "Unified architecture for divide and conquer based tridiagonal system solvers", *IEEE Transactions on Computers*, vol. 43, no. 12, pp. 1413-1425, 1994.

[MaTo90]   S.Martello, P.Toth, *Knapsack Problems: Algorithms and Computer Implementation*. John Wiley and Sons, 1990.

[MaRu80]   T.H.Matheiss, D.S.Rubin, "A survey and comparison of methods for finding all vertices of convex polyhedral sets", *Mathematics of Operations Research*, vol. 5, pp. 167-185, 1967.

[Mc-et-al89] J.McCanny, J.G.McWhirter, E.Swartzlander (eds.), *Systolic Array Processors*, Prentice Hall, 1989.

[McE86]   K.McEvoy, "A formal model for the hierarchical design of synchronous and systolic algorithms", *The University of Leeds, Centre for Theoretical Computer Science, Report*, no. 7.86, 1986.

[McW89]   J.G.McWhirter, "Algorithmic engineering - an emerging discipline", *SPIE, Advanced Algorithms and Architectures for Signal Processing IV*, vol. 1152, 1989.

[Meg90]   G.M.Megson, "Systolic helix for matrix triangularisation with partial pivoting", *Parallel Computing*, vol. 14, pp. 199-206, 1990.

[Meg91]   G.M.Megson, "Automatic systolic algorithm design I: basic synthesis technique", *The University of Newcastle upon Tyne, Computing Science, Technical Report Series*, no. 363, 1991.

[Meg91b]   G.M.Megson, "Automatic systolic algorithm design II: a practical approach", *The University of Newcastle upon Tyne, Computing Science, Technical Report Series*, no. 364, 1991.

248   BIBLIOGRAPHY

[Meg92]   G.M.Megson. *An introduction to systolic algorithm design.* Ox-
             ford Science Publications, Oxford University Press,1992.

[Meg92b]   G.M.Megson, "Mapping a class of run-time dependencies onto
             regular arrays", *The University of Newcastle upon Tyne, Com-
             puting Science, Technical Report Series*, no. 397, 1992.

[Meg93]   G.M.Megson, "Mapping a class of run-time dependencies onto
             regular arrays", *Proceedings 7th International Parallel Process-
             ing Symposium*, Newport Beach, USA, pp. 97-104, IEEE Com-
             puter Society Press, 1993.

[Meg93b]   G.M.Megson, "Mapping certain non-linear dependencies onto
             regular arrays", *The University of Newcastle upon Tyne, Com-
             puting Science, Technical Report Series*, no. 421, 1993.

[Meg94]   G.M.Megson (ed.), *Transformational approaches to systolic de-
             sign.* Parallel and Distributed Computing, Chapman & Hall,
             1994.

[MeCh94]   G.M.Megson, X.Chen, "Partitioning and Mapping for Lower
             Dimensional Given Arrays". *Proceedings 2nd Euromicro Work-
             shop on Parallel and Distributed Computing*, pp. 149-156, IEEE
             Computer Society Press, 1994.

[MeCo91]   G.M.Megson, D.Comish, "Systolic algorithm design environ-
             ments", *Proceedings 2nd International Specialist Seminar on
             Parallel Digital Processors*, pp. 100-104, 1991.

[MeCo92]   G.M.Megson, D.Comish, "Automatic Derivation of Systolic Al-
             gorithms for Kalman Filtering", *Proceedings 3rd IMA Confer-
             ence on Mathematics in Signal Processing*, Warwick, 1992.

[MeCo94]   G.M.Megson, D.Comish, "Systolic algorithm design environ-
             ments (SADEs)", in [Meg94], pp. 205-241, 1994.

[MeCo94b]   G.M.Megson, D.Comish, "Automatic derivation of systolic algo-
             rithms for Kalman filtering", *Mathematics in Signal Processing
             III*, J.C.McWhirter (ed.), Oxford University Press, 1994.

[MeRa95]   G.M.Megson, L.Rapanotti, "Regularising transformations for integral dependencies", *Parallel Algorithms for Irregular Problems: State of the Art*, A. Ferreira, J.D.P. Rolim (eds.), Proceedings, Irregular '94, Summer School and Workshop, University of Geneva, Geneva (Switzerland), August 1994, Kluwer Academic Publishers, 1995.

[Me-et-al95] G.M.Megson, L.Rapanotti and X.Chen, "Automatic Synthesis of Parallel Algorithms", *Solving Combinatorial Optimization Problems in Parallel*, A.Ferreira, P.M.Pardalos (eds.), Special Issue, Lecture Notes in Computer Science, Springer-Verlag, 1995.

[Me-et-al95b] G.M.Megson, L.Rapanotti, G.A.Hedayat, M.F.P.O'Boyle, Z.Chamski, "Detecting and extracting parallelism: old result and a new perspective", *REFLEX Technical Report*, in preparation.

[MeRh84]   R.G.Melhem, W.C.Rheinboldt, "A mathematical model for the verification of systolic networks", *SIAM Journal on Computing*, vol. 13, no. 3, pp. 541-565, 1984.

[MiWi84]   W.L.Miranker, A.Winkler, "Spacetime representations of computational structures", *Computing*, vol. 32, pp. 93-114, 1984.

[Mod88]   J.J.Modi, *Parallel algorithms and matrix computation*. Oxford Applied Mathematics and Computing Science Series Clarendon Press, Oxford, 1988.

[Mol83]   D.I.Moldovan, "On the design of algorithms for VLSI systolic arrays", *Proceedings IEEE*, vol. 71, no. 1, pp. 113-120, 1983.

[Mol87]   D.I.Moldovan, "ADVIS: A Software Package for the Design of Systolic Arrays", *IEEE Transactions on Computer-Aided Design*, vol. CAD-6, no. 1, pp. 33-40, 1987.

[MoFo86]   D.I.Moldovan, A.B.Fortes, "Partitioning and mapping algorithms into fixed size systolic arrays", *IEEE Transactions on Computer*, vol. c-35, no. 1, pp. 1-12, 1986.

[MoCa95]   M.Moonen, F.Catthoor (eds.), *Algorithms and Parallel VLSI Architectures III*. Proceedings International Workshop, Elsevier Science Publisher, 1995.

[Mo-et-al87] W.Moore, A.McCabe, R.Urquhart (eds.), *Systolic Arrays - Papers presented at the 1st International Workshop on Systolic Arrays, Oxford, July 1986*, Adam Hilger, 1987.

[Ner63] E.D.Nering, *Linear algebra and matrix theory.* J.Wiley & Sons Inc., 1963.

[OBo93] M.F.P.O'Boyle. "Program and data transformations for efficient execution on distributed memory architectures", *University of Manchester, PhD Thesis, Technical Report Series*, no. UMCS-93-1-6, 1993.

[Pla99] T.P.Plaks. *Piecewise Regular Arrays.* Gordon and Breach Science Publisher, Parallel Processing Series, 1999.

[PrSh85] F.P.Preparata, M.I.Shamos, *Computational Geometry.* Texts and Monographs in Computer Science, Springer-Verlag, 1985 (Corrected and Expanded Second Printing, 1988).

[PrLi88] D.K.Probst, H.F.Li, "Abstract specification of synchronous data types for VLSI and proving the correctness of systolic network implementations", *IEEE Transactions on Computers*, vol. 37, no. 6, pp. 710-720, 1988.

[Qui83] P.Quinton, "The systematic design of systolic arrays", *Technical Report, Institut National de Recherche en Informatique et en Automatique (INRIA)*, no. 216, 1983.

[Qui84] P.Quinton, "Automatic synthesis of systolic arrays from uniform recurrent equations", *IEEE/ACM, Proceedings 11th Annual International Symposium on Computer Architecture*, 1984.

[QuRo91] P.Quinton, Y.Robert, *Systolic algorithms and architectures.* Masson and Prentice Hall International, 1991.

[QuVa89] P.Quinton, V.Van Dongen, "The mapping of Linear Recurrence Equations on Regular Arrays", *Journal of VLSI Signal Processing*, vol. 1, pp. 95-113, 1989.

[Raj89] S.V.Rajopadhye, "Synthesizing systolic arrays with control signals form recurrence equations", *Distributed Computing*, vol. 3, pp. 88-105, 1989.

[Raj90]   S.V.Rajopadhye, "Algebraic transformations in systolic array synthesis: a case study", *Formal VLSI Specification and Synthesis, VLSI Design Methods I*, pp. 361-370, Elsevier Science Publishers, 1990.

[Raj93]   S.Rajopadhye, "An improved systolic algorithm for the algebraic path problem", *Integration, The VLSI Journal*, vol. 14, pp. 279-296, 1993.

[RaFu87]   S.V.Rajopadhye, R.M.Fujimoto, "Systolic array synthesis by static analysis of program dependencies", *PARLE - Parallel Architectures and Languages Europe, Eindhoven, The Netherlands, June 1987*, Proceedings, Lecture Notes in Computer Science, no. 258, Springer-Verlag, 1987.

[RaFu89]   S.V.Rajopadhye, and R.M.Fujimoto, "Automating systolic array design", *Integration - The VLSI Journal*, vol. 9, pp. 225-242, 1989.

[RaFu90]   S.V.Rajopadhye, R.M.Fujimoto, "Synthesizing systolic arrays from recurrence equations", *Parallel Computing*, vol. 14, no. 2, pp. 163-189, 1990.

[Rao85]   S.K.Rao, *Regular iterative algorithms and their implementations on processor arrays*. Standford University, PhD Thesis, October 1985.

[Ra-Ka88]   S.K.Rao, T.Kailath, "Regular iterative algorithms and their implementations on processor arrays", *Proceedings of IEEE*, vol. 76, no. 3, pp. 259-282, 1988.

[RaMe93]   L.Rapanotti, G.M.Megson, "Pre-Processing in **SADE**: Stage I", *The University of Newcastle upon Tyne, Computing Science, Technical Report Series*, no. 431, 1993.

[RaMe93b]   L.Rapanotti, G.M.Megson. "Pre-Processing in **SADE**: Stage II", *The University of Newcastle upon Tyne, Computing Science, Technical Report Series*, no. 446, 1993.

[RaMe94]   L.Rapanotti, G.M.Megson. "Pre-Processing in **SADE**: Stage III", *The University of Newcastle upon Tyne, Computing Science, Technical Report Series*, no. 471, 1994.

[RaMe94b]   L.Rapanotti, G.M.Megson, "Uniformisation techniques for integral recurrence equations", *The University of Newcastle upon Tyne, Computing Science, Technical Report Series*, no. 478, 1994.

[RaMe94c]   L.Rapanotti, G.M.Megson, "Mapping integral recurrences onto regular arrays", *The University of Newcastle upon Tyne, Computing Science, Technical Report Series*, no. 492, 1994.

[RaMe95]   L.Rapanotti, G.M.Megson, "Uniformisation techniques for reducible integral recurrence equations", in [MoCa95], pp. 283-295, 1995.

[RaMe95b]   L.Rapanotti, G.M.Megson, "A class of dynamic data dependencies and their localisation", *The University of Newcastle upon Tyne, Computing Science, Technical Report Series*, to appear 1995.

[Roc70]   R.T.Rockafellar, *Convex analysis*. Princeton University Press, 1970.

[Rot85]   G.Rote, "A systolic array algorithm for the algebraic path problem (shortest path; matrix inversion)", *Computing*, vol. 34, pp. 191-219, 1985.

[Sa-et-al93]   G.Saghi, H.J.Siegel, J.L.Gray, "Mapping onto three classes of parallel machines: a case study using the cyclic reduction algorithm", *IEEE Proceedings 7th International Parallel Processing Symposium*, Newport Beach, California, 1993.

[SaQu90]   Y.Saouter, P.Quinton, "Computability of recurrence equations", *IRISA, Publication Interne*, no. 521, 1990.

[Sch86]   A.Schrijver, *Theory of linear and integer programming*. Wiley-Interscience Series in Discrete Mathematics and Optimization, 1986.

[ShFo92]   W.Shang, J.A.B.Fortes, "On time mapping of uniform dependence algorithms into lower dimensional processor arrays", *IEEE Transactions on Parallel and Distributed Systems*, vol. 3, no. 3, pp. 350-363, 1992.

[ShFo92b] W.Shang, J.A.B.Fortes, "Independent partitioning of algorithms with uniform dependencies", *IEEE Transactions on Parallel and Distributed Systems*, vol. 41, no. 2, pp. 190-206, 1992.

[Sha87] E.Shapiro (ed.), *Concurrent Prolog*. MIT Press, vol. 1, 1987.

[Sn-et-al85] L.Snyder, L.H.Jamieson, D.B.Gannon, H.J.Siegel (eds.), *Algorithmically specialized parallel computers*, Academic Press, 1985.

[StWi70] J.Stoer, C.Witzgall, *Convexity and Optimization in Finite Dimensions I*. Springer-Verlag, 1970.

[Sw-et-al94] M.F.X.B.Van Swaaij, F.H.M.Franssen, F.V.M.Catthoor, H.J.De Man, "Modelling data and control flow for DSP system synthesis", *VLSI Design Methodologies for Digital Signal Processing Architectures*, M.A.Bayoumi, Kluwer Academic Publishers, 1994.

[Swa85] G.Swart, "Finding the convex hull facet by facet", *Journal of Algorithms*, vol. 6, pp. 7-48, 1985.

[TeTh91] J.Teich, L.Thiele, "Control generation in the design of processor arrays", *Journal of VLSI Signal Processing*, vol. 3, no. 1/2, pp. 77-92, 1991.

[ThTu88] B.C.Thompson, J.V.Tucker, "Synchronous concurrent algorithms", *The University of Leeds, Centre for Theoretical Computer Science, Report*, 1988.

[ThTu91] B.C.Thompson, J.V.Tucker, "Equational specification of synchronous concurrent algorithms and architectures", *University College of Swansea, Department of Mathematics and Computer Science, Computer Science Division, Report Series*, no. CSR 9-91, 1991.

[ThTu94] B.C.Thompson, J.V.Tucker, "Equational specification of synchronous concurrent algorithms and architectures (second edition)", *University College of Swansea, Department of Mathematics and Computer Science, Computer Science Division, Report Series*, no. CSR 15-94, 1994.

[VanD88]   V.Van Dongen, "PRESAGE, a tool for the design of low-cost systolic circuits", *Proceedings IEEE International Symposium on Circuits and Systems*, Espoo, Finland, June 1988.

[Veh94]   U.Vehlies, "DECOMP - A program for mapping DSP algorithms onto systolic arrays", [Meg94], pp. 159-179, 1994.

[VeCr91]   U.Vehlies, A.Crimi, "A compiler for generating dependence graphs of DSP algorithms", *Algorithms and Parallel VLSI Architectures, Volume B: Proceedings*, Elsevier Science Publishers, pp. 319-328, 1991

[VonN66]   J.Von Neumann, *The Theory of Self-Reproducing Automata*. A.W.Burks (ed.), University of Illinois Press, 1966.

[Wil90]   S.A.Williams, *Programming models for parallel systems*. Series in Parallel Computing, Wiley, 1990.

[Wol89]   M.Wolfe, *Optimizing supercompilers for supercomputers*. Research Monographs in Parallel and Distributed Computing, Pitman, 1989.

[Wol86]   S.Wolfram, *Theory and Applications of Cellular Automata*. World Scientific, 1986.

[Wol88]   S.Wolfram, *Mathematica. A System for Doing Mathematics by Computer*. Addison-Wesley, 1988.

[WoDe92]   Y.Wong, J.M.Delosme, "Transformation of broadcasts to propagations in systolic algorithms", *Journal of Parallel and Distributed Computing*, vol. 14, no. 2, pp. 121-145, 1992.

[WoDe92b]   Y.Wong, J.M.Delosme, "Optimisation of computation time for systolic arrays", *IEEE Transactions on Computers*, vol. 41, no. 2, pp. 159-177, 1992.

[Xue92]   J.Xue, *The formal synthesis of control signals for systolic arrays*. The University of Edinburgh, PhD Thesis, no. CST-90-92, April 1992.

[XuLe92]   J.Xue, C.Lengauer, "The synthesis of control signals for one-dimensional systolic arrays", *Integration - The VLSI Journal*, vol. 14, no. 1 pp. 1-32, 1992.

[YaCa88]   Y.Yaacoby, P.R.Cappello, "Scheduling a system of affine recurrence equations onto a systolic array", *Proceedings of the International Conference on Systolic Arrays: Design Methodology and Tools - I*, pp. 373-382, IEEE Press, 1988.

[Zom95]   A.Y.Zomaya (editor), *Parallel and distributed computing handbook*. McGraw-Hill Publishing Company, 1995.

Index